IF YOU POISON US

URANIUM AND NATIVE AMERICANS

PETER H. EICHSTAEDT

IF YOU POISON US

First Edition

Manufactured in the United States of America

Photography by Murrae Haynes

Cover and Text Design by David Skolkin

Library of Congress Cataloging-in-Publication Data

Eichstaedt, Peter H.
If you poison us: uranium and Native Americans
Peter H. Eichstaedt. — 1st ed.
p. cm.
Includes bibliographical references (p. 251) and index.
ISBN 1-878610-40-6
1. Navajo Indians—Claims. 2. Navajo Indians—Government relations. 3. Uranium mines and mining—Southwest, New—History. 4. Radioactivity—Southwest, New—Physiological effect. 5. Navajo Indians—Social conditions. I. Title.
E99.N3E29 1994
363.11'962234932'089972—dc20 94-7306
CIP

Red Crane Books
2008 Rosina Street, Suite B
Santa Fe, New Mexico 87505

To the Dineh *(Navajo) and all Native Americans*
who have endured, for their time
will come again

Contents

Foreword

THE HISTORY OF OUR NATION'S relations with American Indians is one of ignorance, indifference, exploitation, and broken promises. When land occupied by the Indians was needed by settlers, or for some other public purpose, it was seized and the Indians herded onto apparently barren reservations. Then, when these reservation lands turned out to be rich in minerals and other resources, they were leased to mining companies, ranchers, and others, with little or no regard for the rights of the native inhabitants, their livelihood, or the long-term effects on the land. Often, only token payments were made for these extractive uses and sometimes none at all because the secretary of the interior, the designated federal trustee, failed to ensure payment.

Over and over again, the interests of this country's native population have been sacrificed to the needs of the dominant society in clear violation of the federal trust responsibility established by solemn treaties and federal statutes. Peter Eichstaedt's remarkable book provides an objective and balanced account of one of the most shameful chapters in this shameful history.

It is ironic that, despite the disgraceful way we have treated them, Indians have remained loyal and patriotic citizens. Late in World War II, a secure uranium supply became critical to producing an atomic bomb. Some of the richest mines were found on Indian lands on the Colorado Plateau—the Navajo Reservation and the lands of Laguna Pueblo. The U.S. government appealed to the Indians to help defend the country, and the tribes responded favorably. The mines would also provide much-needed income for a people whose existence was often precarious. However, neither the tribal governments nor the miners were told that there were hidden costs—that the mines were dangerous.

By the start of the Manhattan Project, there was already good evidence that working with uranium and radium could cause lung disease, cancer, and other fatal health problems. As the Public Health Service began to study miners on the Colorado Plateau, about a quarter of whom were Indians, there was no longer any doubt that uranium mining was an extremely hazardous operation that required protection for the miners. Yet, the Atomic Energy Commission, for years the sole purchaser of the uranium, consistently maintained that it had no responsibility for conditions in the mines, and mining companies refused to take the most elementary safety measures. Nor did the secretary of the interior take any effective action to protect them. Then, when death started to come to the miners, slowly and painfully, years later, mining companies and federal and state agencies argued over who should compensate the victims and their families—or even if any compensation was due.

The Navajos spent the 1970s and 1980s fighting in Congress and the courts to obtain some help for the families of sick and dying miners, and where they did not encounter indifference, they were fought by government officials. Some of this was undoubtedly the result of ignorance and the natural tendency of bureaucrats to deny the existence of problems they created. Some of it, however, was the result of a cynical disregard of the rights and welfare of the Indians. After the Navajos made fruitless appeals to Congress and after nearly a decade of litigation, the government was able to escape legal liability under the Federal Tort Claims Act through a loophole known as the "discretionary function" exception—which insulates certain forms of governmental conduct from court review.

However, the litigation had an unexpected benefit. The evidence collected for the court cases showed, beyond reasonable debate, that compensation was due the miners, and even the court decisions indicated that legislative relief was warranted. With court relief no longer possible, the Navajos went back to Congress again and finally, in 1990, convinced Congress to enact remedial legislation through the Radiation Exposure Compensation Act. However, when this legislation was finally signed into law, its provisions, designed for a white world, took little account of the realities of life on the reservation or the limitations of medical facilities there.

One of the political trade-offs of getting a compensation bill passed was allowing the Department of Justice to administer it. This was not unlike putting the fox in charge of the chicken coop since the justice department had not only bitterly fought all attempts to get such a bill passed but had also fought the earlier claims in court. As administered by Department of Justice lawyers, the documentation that has been required for claims is unavailable or extremely difficult for traditional Navajos to obtain. For them, the process of obtaining compensation has been excruciating and frustrating. The sick miners, or their widows, have had to travel great distances on primitive roads to meet the unending demands of Washington for work records or marriage certificates that often do not exist.

The effects of the mining have been especially devastating for the Navajos because the deaths have been concentrated in small communities and because so many of them live in extreme poverty. Often when a miner dies, his family is left destitute. What is particularly outrageous about the situation is that while many miners and their families have suffered damage, the unduly strict requirements of the statute have disqualified many potential Navajo claimants. The total number of claims they have actually filed so far is less than 500. Yet, in the supposed interest of protecting the U.S. treasury from fraudulent or invalid claims, the justice department's lawyers have insisted on the strictest possible interpretation of the law in processing those relatively few Navajo claims, even where the result has been to subvert Congress's clear intent to provide compassionate compensation. While, as shown in the epilogue, the pace of processing has very recently improved, it is still true that, compared with the total number of miners, only a small number of Navajo claimants have been successful.

The damage of the uranium years has not been limited to the miners and their families and friends. When the boom was over and the mining companies pulled out, they left behind enormous piles of radioactive mine waste from milling operations and more than a thousand open mines on Indian lands. There has been no federal effort to clean up these mines, many of which leak deadly gases and spread radioactive dust—poison that is no less deadly than the smallpox-infected blankets that the government issued to thousands of Mandan Indians in the nineteenth century. While the Indians themselves have started reclamation, with what little money they can scrape together, far more is needed to provide effective relief. Having poisoned the people and the land, the mining companies and the government have left the Indians to fend for themselves.

If we poison Indian lands, we do far more harm than the damage we cause to individuals who suffer the immediate consequences. It must be remembered that Indian tribes are sovereign nations defined, in large part, by the lands they occupy. Indian tribes cannot relocate their sovereign power to new territory if their existing territory is made uninhabitable. If we destroy tribal territory, we strike at the very roots of tribal existence.

Those of us who work in the field of Indian law like to think that there has been some progress in the last part of this century in vindicating Indian needs and rights. However, whatever progress has been made cannot obscure the fact that Indians are still the losers in any clash of our nation's public policy with the rights of tribes and their individual members. This book makes a powerful case that something more must be done to remedy the evil results of the government's uranium policy—results that have severely damaged the land as well as the miners and the families who live on it. Surely, as a great and powerful nation, we can do better than we have done so far.

Jerry C. Straus
Washington, D.C.
April 1994

Acknowledgments

THIS BOOK WOULD NOT have been possible without extensive help and support from many people. They are:

Michael and Marianne O'Shaughnessy of Red Crane Books, whose dedication to Native Americans sparked this project; Carol Caruthers of Red Crane Books for direction, advice, and encouragement; Murrae Haynes, who shares a great concern for Native Americans, and whose photography and sense of adventure made for good road trips; Perry Charley, of the Navajo Abandoned Mine Lands Reclamation office, who guided me in many ways and instructed me in Navajo culture and language; Phillip Harrison, Jr., whose dedication to the cause of the Navajo uranium miners and millers is an inspiration to all who know him and whose tireless help greatly enhanced this book; William Chenoweth, of Grand Junction, Colorado, who was an invaluable source of information and documents on the history of uranium production on the Navajo lands and who supported and participated in this project beyond all expectations; Stewart Udall, whose dedication to justice deserves a book of its own, and who provided me with moral and spiritual guidance as well as many valuable source documents; Paul Robinson and Chris Shuey of the Southwest Research and Information Center in Albuquerque, who provided insights, inspiration, and documentation; Harry Tome, who granted me access to his personal files and provided detailed accounts of the Navajos' struggle for justice; Timothy Hugh-Benally and Dr. Louise Abel of the Office of Navajo

Uranium Workers and the Indian Health Service in Shiprock, New Mexico, for the information they provided and their helpful critique of the manuscript; Ray Tsingine of the Navajo Abandoned Mine Lands Reclamation office in Tuba City for his helpful guidance and information; Chris Norvell for skillful piloting, and the friendly, dedicated people at LightHawk, the environmental air force, for their support and cooperation in obtaining aerial photographs; Jay Lazarus, geologist and friend, who provided necessary documents and explanations; Russel Edge of the Department of Energy for his gracious cooperation and explanations; all the Navajo miners and millers who shared their stories and agreed to be interviewed and photographed for the book; Drs. Jonathan Samet and David James of Albuquerque and Dr. Victor Archer of Salt Lake City for their expertise and insightful comments on the manuscript; Mrs. Irene Wade of California, who provided her late husband's book and other documents about the early days; Manuel Pino of Acoma Pueblo and Alvino Waconda of Laguna Pueblo for their help, and in hope of justice for their communities; most of all, I must acknowledge the support and love I received from my wife Connie, and children Matthew and Ashley, while I was researching or writing; and, of course, Robert Parker for his encouragement.

Introduction

THIS BOOK IS THE STORY OF THE SACRIFICES OF LIVES, families, and land that Native Americans in the Southwest have made in America's quest for nuclear superiority.

The race to build an atomic weapon that began in secrecy during World War II, and later, the atomic arsenal of the Cold War, required a steady supply of uranium ore. That supply was found in the Four Corners area of the Southwest, the heart of Indian country and the Colorado Plateau. The U.S. government ensured uranium production by giving the mining companies large financial incentives, by keeping the uranium miners uninformed about the dangers of the mines, and by appealing to the patriotic spirit of the miners, saying that the nation's security was at stake.

About a quarter of the miners were Native Americans, mostly Navajos, who labored in the uranium mines and processing mills. They dug the uranium ore with pick and shovel in the small mines called "dog holes" or blasted the rocks with dynamite, breathing radon gas and silica-laden dust. They ate food tainted with uranium oxide and drank the contaminated water that dripped from the mine walls. They carried uranium home to their wives and children on their shoes, clothes, and bodies. Yet, in the face of mounting evidence and warnings from Public Health Service physicians, mining companies and government agencies refused to acknowledge that there was danger—that the miners were harvesting their own death.

Nearly two decades after the mining began and only after deaths began to mount among the miners did the government impose radiation exposure standards on the uranium mines—in spite of relentless opposition from mining companies.

Another two decades passed and more deaths occurred before the U.S. Congress finally acknowledged in 1990 that the hapless victims of the nuclear age—the miners, atomic fallout victims, and the military's human guinea pigs—should be compensated for their losses.

In the Navajo perspective of the world and the universe, man is just a small part of a much larger whole, a system in which everything exists in balance: the earth, the sky, the spirits, and all living creatures. To maintain and preserve this harmony is to walk in beauty. When people, plants, or animals get sick, it is usually because the harmony has been disturbed. Navajo medicine men perform healing ceremonies to restore harmony so that a patient can once again walk in beauty.

However, there are no healing ceremonies that can restore the harmony that has been disturbed on Indian lands in the Southwest. No Navajo ceremonies and none of the white man's medicine can restore life to the men who died mining and milling uranium.

Today Native Americans continue to reap a bitter harvest for their patriotic role in World War II and the Cold War. Undetermined tons of exposed radioactive mine waste remain on native lands. Rainwater has leached uranium byproducts and toxic metals into underground water, with potentially long-lasting consequences. Small uranium pit mines remain open, filled with water, inviting children to swim and animals to drink. At Laguna Pueblo, an open-pit mine that covers nearly 3,000 acres remained untouched for seven years after operations stopped, until the pueblo itself started reclamation.

This book is the story of how uranium mining began on Indian lands in the American West, how it was conducted, and how its deadly legacy still lingers in the lives of the men, women, and children whose harmony and homelands have been destroyed.

IF YOU POISON US

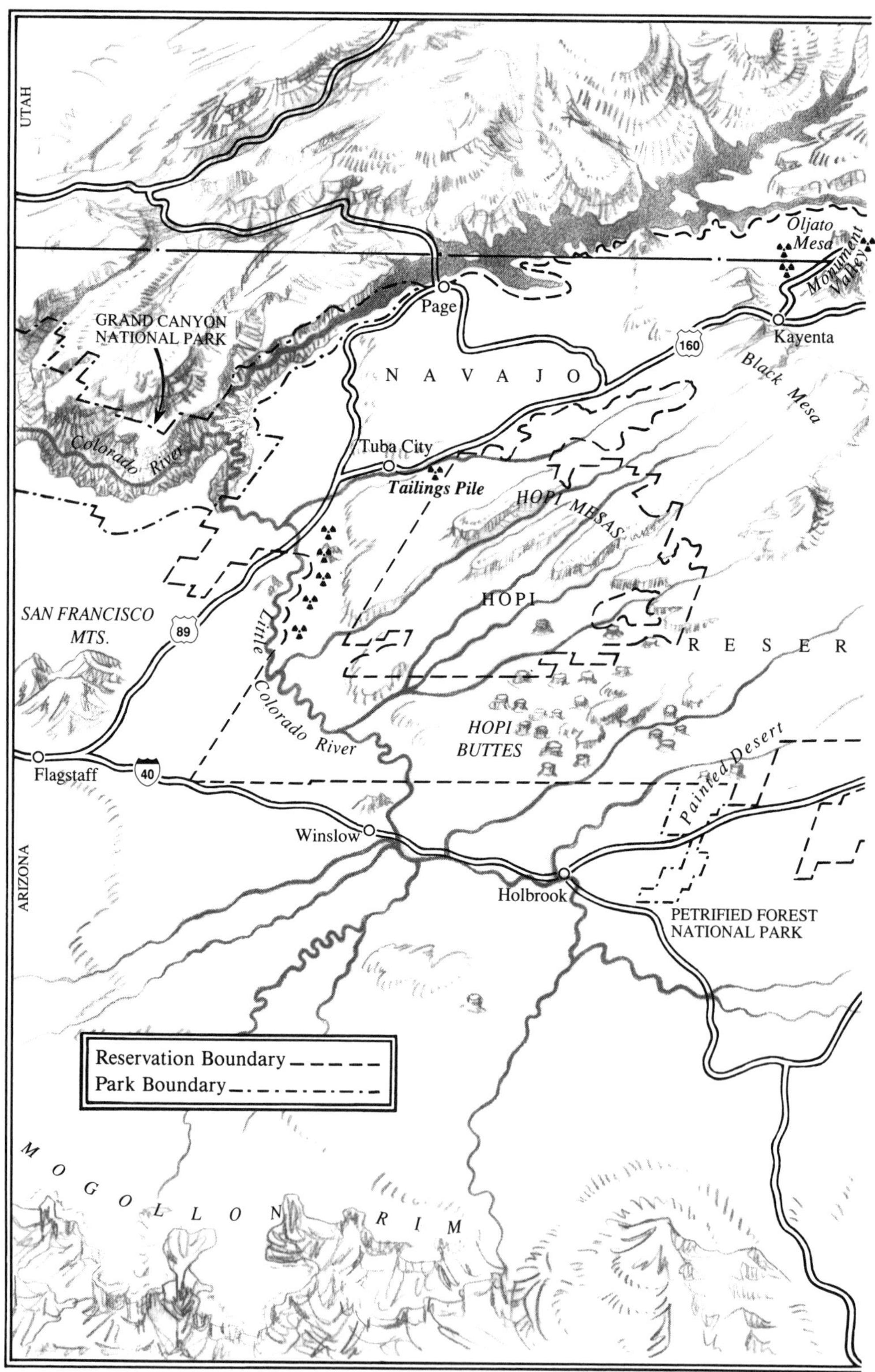

UTAH
ARIZONA
GRAND CANYON NATIONAL PARK
Colorado River
Page
Oljato Mesa
Monument Valley
Kayenta
160
Black Mesa
N A V A J O
Tuba City
Tailings Pile
HOPI MESAS
HOPI
R E S E R
SAN FRANCISCO MTS.
89
Little Colorado River
HOPI BUTTES
Painted Desert
Flagstaff
40
Winslow
Holbrook
PETRIFIED FOREST NATIONAL PARK
Reservation Boundary
Park Boundary
M O G O L L O N R I M

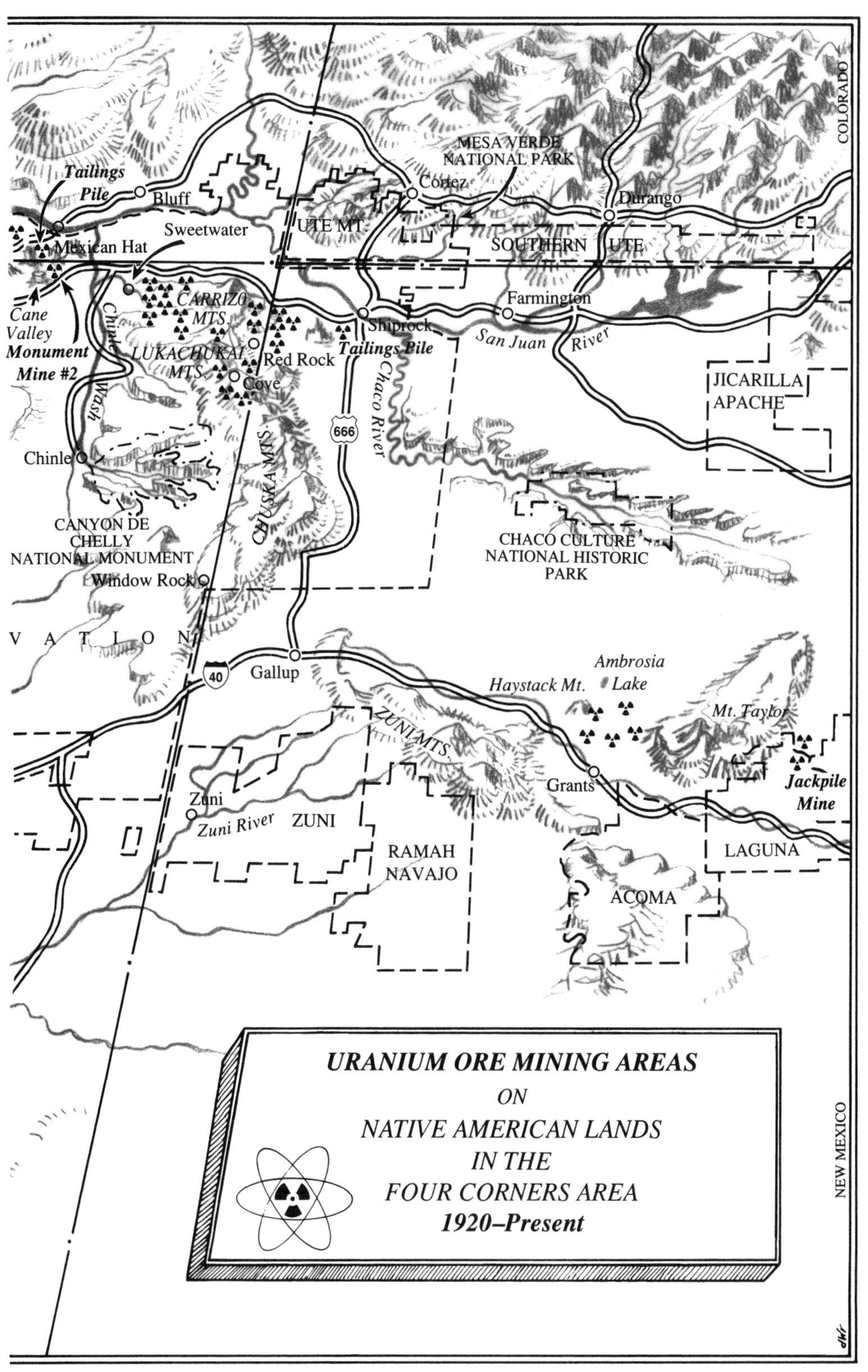

COLORADO
MESA VERDE
NATIONAL PARK
Tailings
Pile
Bluff
Cortez
Durango
Sweetwater
UTE MT.
Mexican Hat
SOUTHERN UTE
Farmington
CARRIZO
MTS.
Cane
Valley
Monument
Mine #2
Chinle
LUKACHUKAI
MTS.
Red Rock
Cove
Shiprock
Tailings Pile
San Juan River
Chaco River
666
Wash
JICARILLA
APACHE
CHUSKA MTS.
Chinle
CANYON DE
CHELLY
NATIONAL MONUMENT
CHACO CULTURE
NATIONAL HISTORIC
PARK
Window Rock
V A T I O N
40
Gallup
Ambrosia
Lake
Haystack Mt.
Mt. Taylor
ZUNI MTS.
Grants
Jackpile
Mine
Zuni
Zuni River
ZUNI
RAMAH
NAVAJO
LAGUNA
ACOMA
URANIUM ORE MINING AREAS
ON
NATIVE AMERICAN LANDS
IN THE
FOUR CORNERS AREA
1920–Present
NEW MEXICO

1 The Shadow of the Red Rock

On a clear, cold day in early February 1993, two dozen people gathered in a modest prefabricated building adjacent to the Indian Health Service (IHS) hospital in Shiprock, New Mexico, a community on the eastern edge of the Navajo Reservation. The group assembled there was a mixture of medical doctors, lawyers, government officials, and news reporters. It included Helene Goldberg, tort branch director and compensation program administrator with the U.S. Department of Justice, and Stewart Udall, the former secretary of the interior and congressman, a lifelong advocate for Native Americans. Excitement and sadness were in the air.

The results of a comprehensive attempt to identify Navajos who at one time or another had worked in uranium mines and mills of the Southwest were about to be announced. The statistics had been compiled by Timothy Hugh-Benally, a Navajo and the head of the Office of Navajo Uranium Workers, which was established by the Navajo tribe after the Radiation Exposure Compensation Act of 1990 (RECA) was signed into law. The act provided $100,000 in "compassion payments" to uranium miners who were diagnosed with cancer or other respiratory ailments linked to uranium mining, or to the families of uranium miners who had died.

Collecting the statistics had been tedious, time-consuming work. Health officials had to visit each of the many chapter houses across the 25,000-square-mile Navajo Reservation to personally interview former miners and millers, many of whom did not speak English. The miners were asked about their work history and health. Five hundred and forty-nine were brought to the Office of Navajo Uranium Workers for medical examinations and tests. They were given chest X-rays and breathing tests to determine damage to their lung capacity. The statistics were both confusing and alarming. Four of the miners were found to have lung cancer. However, 10 of the 549 had died of unknown causes after the screening began, and 1 miner had a heart attack while in the office. The tests showed that under the guidelines drawn up by the Department of Justice lawyers, only 5 of the 549 miners were eligible for compensation payments for respiratory disease. The doctors and lawyers who worked with the Navajo victims were irate. They saw daily the death, disease, and destruction that had been delivered to the Navajo people and their lands, but the tests did not reflect the obvious.[1]

Dr. Louise Abel, a compassionate woman with a warm and friendly smile who works with the Indian Health Service in Shiprock, stood to address the crowd. "Every human being reacts differently" when exposed to radiation, she said. Some of the miners and millers were exposed to huge amounts of radiation over their working careers. For some, "It [exposure] had no effect on them." Others died even though they had received only relatively small doses in four or five years of uranium mining. "I feel the ones who had the worst cases [of cancer] are dead," she told the group. "The ones who were the most affected have died. What we're seeing are the survivors."

Albert Tinhorn, a Navajo chapter leader from the Dennehotso area of Arizona, a stout man who sports a large black hat with a bright feather, stood to speak in a deep and passionate voice. "Everybody is just like dying off, and they need to be compensated," he said.

The compensation authorized by Congress, however, had been excruciatingly slow in coming. The Navajos were justifiably bitter. As Hugh-Benally put it, "We provided the uranium to the U.S. government, and [it] did not tell us it was dangerous. And now the government tells us you have to die to get compensated.... Congress said it didn't know about the side effects of uranium.... They knew what the effects were. The people feel like they were a study project. The early studies prove they [the government] knew of the effects."

The importance of the meeting in Shiprock can easily be lost without an understanding of how uranium came to be mined on the Navajo Reservation and other Indian lands in the Southwest. It is a saga of greed and indifference that starts in a small laboratory in Germany and spans the twentieth century.

* * *

Uranium has been around since the beginning of time, and it most likely will be around until the end of time. However, this is not the human time that we all know, the time of ticking clocks, the time of digital displays, the time of

MEETING AT THE SHIPROCK OFFICE OF NAVAJO URANIUM WORKERS ON FEBRUARY 10, 1993. LEFT TO RIGHT, JANE COREY, PUBLIC HEALTH SERVICE NURSE, DR. LOUISE ABEL, AND TIMOTHY HUGH-BENALLY.
PHOTO: AL CABRAL

NAVAJO RESERVATION NEAR MEXICAN HAT, UTAH.
PHOTO: PETER EICHSTAEDT

dropping kids off at school, the time of airplane departures and arrivals. This is geologic time, the time of rocks. It is measured in hundreds of millions of years, ancient epochs that extend so far into the past that even genetic memory does not exist. Its days are called Paleozoic, Mesozoic, and Cenozoic. Its hours are called Cambrian, Permian, Triassic, and Jurassic. Its minutes are called Eocene, Oligocene, and Miocene.

Uranium is found all over the world. A knowledgeable geologist can find it almost by scanning the landscape. It is often found in sandstone—the sedimentary rock that was once at the bottom of the oceans and rivers that covered the planet about 300 million years ago. Over enormous spans of geologic time, the crust of the earth convulsed, creating land by forces so powerful that nuclear weapons are mere firecrackers by comparison. These ocean bottoms of compacted sand and rock were lifted and exposed. Another 100 million years of sun, wind, and rain, penetrating frost and ice, and the edges of these thin uranium deposits were revealed.

The existence of uranium in Europe was known by the late 1700s. Found in an old silver and metal mine in the Erzgebirge or Ore Mountains near the town of St. Joachimsthal in what was then Bohemia, the uranium-containing ore was called pitchblende. Uranium oxide was extracted from pitchblende in 1789 and named uranium after the planet Uranus, which had recently been discovered. Until its radioactive properties became known, uranium was used in pottery glazes and iridescent glass. In America, from 1936 to 1943, uranium was still being used to provide a bright red-orange glaze to a line of popular dinnerware known as Fiesta ware.

The radioactive significance of uranium was virtually unknown until 1895 when the German physicist Wilhelm Konrad Roentgen (1845–1923) studied the strange glow that came from a Crooke's tube, a gas-filled vacuum tube. He named the highly penetrating rays "X-rays"; the name was later changed to Roentgen rays. With the work of Antoine-Henri Becquerel and Marie and Pierre Curie, the world soon after learned of the radioactive elements that are contained in uranium ore. The discovery of radioactivity led to the realization that certain elements are unstable and decompose into other elements, such as radium. In the process, rays are emitted and energy released. This revelation startled the scientific world. However, while the possibilities of this new element seemed endless, there was another side to the apparent magic.

Madam Curie had isolated radium as a highly radioactive element that was found in uranium-bearing ores, and in 1901 the Curies gave Becquerel a small vial of relatively pure radium. He placed it in his vest pocket and two weeks later noticed a red, blotchy patch of skin on the side where the vial had been. The redness developed into a sore that took a long time to heal. Becquerel reported the incident to the Curies; Pierre repeated the experiment on himself and confirmed that the burn had come from radium.

Instead of alarm, this dramatic effect brought new hope for a cancer cure, especially for tumors and other cancerlike growths. While living cells die when they are exposed to large amounts of radiation, older, mature cells of tissues are unaffected by controlled, smaller doses. Only young, rapidly dividing cancer cells, such as those found in tumors, are killed. The explanation at the time was that the young cells are more vulnerable. Although Marie Curie died of radium poisoning in 1934, her death did not provide the signal it should have, for there was already a growing body of evidence that radioactive materials were deadly.

In the United States, meanwhile, many deposits of pitchblende had been found and were being worked. By the turn of the century, pitchblende had been found in Connecticut, South Carolina, Texas, the Black Hills of South Dakota, and Gilpin County, Colorado. As early as 1871, pitchblende from the Colorado mines was being shipped to Europe for glazing and glass compounds, and was prized for its high content of uranium oxide.

* * *

The Colorado Plateau is a high and dry expanse of weathered rock shrouded in hues of gray and tan, broken only by rough buttes of rusty reds—the sandstone deposited when the West lay under a large sea. Today this rugged plateau is crisscrossed by the arbitrary and invisible borders of the states of Colorado, Utah, Arizona, and New Mexico. This is the Four Corners area, the only spot in the United States shared by four states. It is marked by a big brass plate screwed into a chunk of flat, red sandstone. It is one of the most photographed regions of the United States. Nearby, Navajos sit patiently under shaded stalls displaying their colorful rugs; handcrafted silver belts, bracelets, and bolos; their pottery and baskets.

From this spot the horizon fades from view, squint as one might, and as one must under the unsympathetic sun. In this rambling, rocky place some of the most dramatic landforms in the world can be found. It is where the sculpting hand of nature has carved colorful canyons and left dramatic, towering buttes and spires that remind us of the insignificance of man. It is a plateau cut by rivers like the San Juan, the Green, the Colorado, and the Little Colorado, each finding its way through and across these ageless rocks.

It was from this plateau that the Navajos were driven about 130 years ago by the famed Indian fighter and American hero Kit Carson. And it is here that they returned, decimated by disease and poverty after the attempt to "civilize" them in the inhospitable and barren land of the Bosque Redondo at Fort Sumner, New Mexico, where the U.S. government set a pattern for the next century and a half of broken promises, unfair treatment, and inability or unwillingness to understand or address the Indians' needs. The land to which the Navajos were finally allowed to return in Arizona, New Mexico, and Utah was only a small part of the area they had claimed as their home. It was on this land that large uranium deposits would be found.

In light of the events of the past few decades, it is ironic that when they returned, the ceremonial sand paintings that are an integral part of the Navajos' spiritual life and healing ceremonies were made from rocks that contained the powerful yellow dust and the gray and black sands that are often found with uranium-bearing ore.

2 A Grave Question of Prosperity

THE BRIGHT YELLOWS in the Colorado Plateau rocks also attracted the attention of the explorer John Wetherill in 1898 when he led an archaeological expedition into San Juan County in southern Utah. While the rest of his party ate lunch in a shady cave and talked about Indian artifacts, Wetherill quietly wrote up a claim and covered the paper with rocks. The site of Wetherill's claim, filed officially by an acquaintance forty-five years later, became one of Utah's richest uranium deposits—the Blue Lizard mine.

Others like Wetherill who spent their days wandering the Colorado Plateau in the late 1800s also knew of the potential value of the yellow, crumbly minerals he had found because word of a potentially valuable new mineral had spread like wildfire. The mineral was carnotite, which contains both uranium and vanadium, a metallic element used to harden steel.

To meet the demand for vanadium and to reduce the cost of shipping crude ore to France and Germany for processing, Charles Poulot and fellow Frenchman Charles Voilleque in 1900 started the first uranium and vanadium processing plant in the United States at the Cashin copper mine in the Paradox Valley of west-central Colorado. Other mines and mills opened over the next several years, but all found limited success. The Poulot mill failed after two years even though it produced fifteen thousand pounds of uranium and vanadium concentrates.[1] The chemicals needed to leach the uranium and vanadium effectively were in short supply and had to be carted across treacherous terrain, and miners were often unable to assay their ore accurately.

As the cancer-killing properties of radium kindled the interest of the medical community in Europe, two American doctors, Robert Abbe, a surgeon at St. Luke's Hospital in New York City, and Howard A. Kelly of Johns Hopkins in Baltimore, were also enthusiastic. In 1904, Kelly had obtained a small amount of radium from the Curies. His initial work with it was so successful that he bought more in 1910, and by 1913 he was convinced of its effectiveness.[2]

Meanwhile, two other figures were infected with enthusiasm about radium: Joseph and Michael Flannery. The brothers founded the American Vanadium Company in Peru, the world's largest supplier of vanadium, but were devastated by the death of their sister from an inoperable form of cancer. They had been frustrated in their attempts to obtain the radium which they were convinced held the best promise for her cure. As a result, the brothers sold their Minasragra mine in Peru and focused on the United States, determined to produce radium from existing deposits of carnotite on the Colorado Plateau. They were confident that the demand for radium would only increase.

In 1910, the Flannerys formed the Standard Chemical Company and bought the biggest and best carnotite claims they could find in Colorado and Utah. They set up the second processing mill in North America at Uravan, Colorado, and produced radium until 1922. The total investment in turn-of-the-century dollars was $650,000. By the end of 1913, the company had produced just two grams of radium, but by 1914 it was ready to produce a gram a month to meet an order from Germany for fourteen grams. At a price of about $150,000 per gram, the possibility for profits looked good.[3]

* * *

While prospectors and entrepreneurs mined carnotite on the Colorado Plateau, the Navajos attempted to pick up the pieces of their lives. Descendants of people of the far north who about a thousand years ago migrated south, the Navajos had roamed the plateau as raiders and hunters, eventually settling down to farm and trade with their Pueblo neighbors. This interchange, however, did not protect the Pueblos from Navajo raids for slaves and livestock.

After the Long Walk to the Bosque Redondo and back, the Navajos returned an apparently reformed people who promised to farm, send their children to the white man's schools, and to stop raiding and fighting. However, Kit Carson had destroyed thousands of acres of fields and orchards that the Navajos had lived on, and weather and insects ruined the crops they attempted to start with the seeds the white men gave them. The sheep and goats given to the destitute Navajos—two for every man, woman, and child—did much better. In a remarkable show of resilience and courage, the Navajos built up their flocks, began to weave rugs for sale, and learned to do the silverwork for which they are now famous.

The old ways were never lost, though, for survival, then as now, depended on an ability to live in harmony with the land, using it with respect. They hunted plentiful rabbits, squirrels, and prairie dogs, and occasional larger wildlife like the deer and antelope found in the high desert. Children were taught to survive in the wilderness at a very early age, as well as to contribute to the well-being of the family. A boy of eight could travel for days across the harsh country wearing nothing but moccasins and a loincloth and carrying only a sharp piece of chipped, black obsidian. He could find the fresh tracks of a rabbit and follow them to its burrow. Breaking off a branch of a sagebrush and shredding it, he would stick it down the hole and twist until it tangled into the rabbit's fur, and then he would pull the rabbit from the hole and kill it with a swift blow from a rock. The hand-sized obsidian was used to gut the rabbit, which could then be carried for hours to the next camp or back home for the family meal.

A tender root quenched thirst in the desert and provided quick energy. A piece of the root was chewed until it became mushy and gooey and then dropped on an anthill and eaten when it was covered with ants. A bow and arrow and the company of a friend were sufficient to obtain a meal of prairie dog. The hunter would walk around the prairie dog hole while the partner remained on the other side. When the curious prairie dog popped up to look in the direction from which the intruders had come, he was drilled by a small arrow from several feet away and pinned so that he could not retreat.

The Navajos learned as children that once game became scarce, it was time to move on. Nothing was hunted to extinction; always enough animals were left to reproduce so that hunting would remain good for another year. Man was just one of the many animals on the earth and depended on other animals for food, clothing, and shelter. Only the white man, the Navajos learned, would take from nature until everything was gone.

Although survival outdoors meant learning the ways of the high desert, much of Navajo life took place inside the hogan, a simple yet efficient, dome-shaped wood and adobe shelter where a fire was kept going and a pot of stewed rabbit or prairie dog was warmed. In the late afternoon, the family gathered for their daily meal. It was a time to talk about the day, the hunt, and the children. Grandfather would tell the children the myths and legends, the ways of the eagle, what the eagle spirit was and what it did for the *Dineh*, the Navajo name for themselves. The family would sit around the fire wrapped in blankets and enjoy the long and leisurely time, the time when a new baby was passed around, made to laugh, and allowed to grow up quickly, but with a sense of belonging to the family and the hogan.[4]

While smoke from the hogans drifted peacefully skyward, a power struggle was being waged nearly two thousand miles away in Washington, D.C., that would in time come right to the door of the Navajos and other Indian tribes scattered over the mineral-rich land of the Southwest.

* * *

Uranium was not the only source of wealth in the West. The country was rich in minerals, timber, and most of all, land. With the slaughter of the vast herds of bison that covered the plains, the waist-high grass became available to cattle. The gold, silver, and other minerals that lay beneath the surface of the earth were sought by the industrialized East, whose insatiable appetite for seemingly endless raw materials was well known to the entrepreneurs that flooded the West. Unfortunately for development interests, the West was not completely open. Still in the way were Native Americans who had populated the continent for thousands of years and who had no written language, no technology, no machinery, no schools, a strange religion, no tradition of individual ownership of land, and no apparent social organization at all beyond the tribe.

What use could these apparently primitive people have for the enormous resources of natural wealth that awaited exploitation? To clear the way, hundreds of thousands of Native Americans were systematically herded onto reservations throughout the West as they were either defeated by the U.S. Army or voluntarily placed their fate in the hands of federal agents, knowing that resistance meant annihilation by overwhelming firepower and manpower. However, no matter where the Native Americans were placed, they seemed to be in the way of exploitation of some resource.

Even if a tribe was astute enough to know that there were profits to be made in controlling a natural resource, what rights, after all, did they have to do so? This was more than a casual question, and it had wide-ranging ramifications. Under American law, which today some tribes still do not formally recognize, one needed to show proof of ownership of land in order to exploit it. The Indians had none. The land, the Indians believed, belonged to no one. Man was simply one of many temporary occupants. The concept of a single person "owning" or buying and selling a piece of land was not only alien and bizarre to the Indians, it was laughable. This attitude rendered the Indians defenseless and vulnerable in the face of federal lawyers at the turn of the century, who argued that the only reason Indians had any land at all was due to the graciousness of the U.S. government. Indians, being considered wards of the state, had no property rights.

This being the case, the government had the legal right to allot the land to the native occupants of the continent when, where, and how it saw fit. In the final analysis, the wants and desires of the Native Americans did not need to be taken into account. Most of all, the Indians were viewed as having no right to lay claim to and derive profits from the natural wealth of the land they occupied. Not everyone agreed, however.

At the turn of the century, an organization called the Indian Rights Association, a national group of Indian rights activists, objected to this argument and forced the issue to be addressed in a congressional hearing on the leasing of Indian lands. Fearing that corrupt Indian agents, the employees of the federal government, were using their powerful positions to take advantage of the Indians on behalf of ranchers and others, the rights group sided with senators who opposed the decision to give the Indians the right to lease their land. The issue became emotional, pitting liberals sympathetic to the plight of Indians against overwhelming expansionist and development interests, specifically those of the ranching, mining, and timber industries. Ironically, the rights group misread the protectionist intentions of the Indian agents—who were arguing that the Indians *did* have the right to lease their land—and played into the hands of the exploitive ranchers, miners, and timber cutters who wanted the resources on Indian lands for their own profit.

In 1902, a hearing was held on Senate Bill 145, introduced in late 1901, which would open the Uintah Reservation to grazing and mineral leases that would be sold to the highest bidders, the money from the sales being placed in trust accounts for the Uintah Indians. The Uintahs had been reduced to a band of about 800 men, women, and children who, since 1875, had lived peacefully in what was known as the Uintah River Valley of northern Utah. Like many Indian lands across the country, these large tracts were coveted by American entrepreneurs. They were prime rangelands in the eyes of cattle barons, who viewed the West as their own private empire, even though many of them had been on the land less than a dozen years. Miners and mineral prospectors viewed these lands as unfairly exempted from exploitation. The West, they believed, was fair game for anyone who had the ability to exploit it.

Aggressive and enterprising miners and ranchers across the West took the initiative and secured proposed grazing and mineral development leases with several tribes, including the Uintahs and the Standing Rock Sioux of South Dakota. These leases were not only secured with the permission of the U.S. Commission on Indian Affairs, the predecessor agency to the Bureau of Indian Affairs, but with the agency's assistance. Called to testify at the Senate hearing, the agency argued vehemently that this was not only legal but that it was being done with the full approval of the Indian tribes and was, after all, for their benefit.

The prevailing attitude that the Indians owned nothing and therefore were not entitled to dictate whether profits were to be earned on the land they did not own is evident in the transcripts of the hearing. William Jones, the commissioner of Indian affairs, set the tone when he said, "I do not believe in reserving large tracts of land for the exclusive use of Indians. I believe it ought to be thrown open as rapidly as possible. I understand, though, that there is a treaty arrangement with those Indians which will make it necessary for you to treat with them before it can be thrown open to settlement…and I think before you get them to agree to open the reservation you have got to use some arbitrary means to open the land."[5]

The land under the control of the Sioux and the Uintahs totaled about 2 million acres. The bill's sponsor, Sen. Joseph Rawlins, argued that the Indians should not be paid directly for the land that might be taken from them because they would squander the money. "If any money is given them they gamble it away and dispense with it. It is not for the best interest of the Indians to pay them money.... These are Indians who can not intelligently deal with this subject independently.... This bill does not take away from them anything. It converts their lands into a fund which will be applied to their benefit."[6]

At the time of the hearing, the Uintahs had already agreed to two grazing leases and a mining lease with the Raven Mining Company. The company was allowed to spend two years exploring reservation lands and then stake a claim for 640 acres that they would work. The tribe was to get only 5 percent of the value of the resources that came off the reservation, which, sadly, is comparable to similar leases tribes would establish for uranium some fifty years later.

Commissioner Jones explained that his agency had little authority over what the Indians did or did not do with their land. They only had the authority to permit people who had business on Indian lands to negotiate deals with Indians. Despite this contention, the Commission on Indian Affairs helped assemble general tribal meetings to approve the leases by the Indian tribes. Although such actions may well have been sympathetic and intended to provide the Indians with at least some profit and control over their own fate, they appeared to be taken on behalf of ranchers and mining companies.

In the case of the grazing leases proposed for the Standing Rock Sioux, the commission directly obtained 771 consent signatures of males over eighteen years of age on the reservation out of a possible 983. The signatures were obtained in lieu of approval by the tribal council because a council did not exist at the time, as was true of most other tribes. In addition, the nearly 1 million acres put up for grazing leases was advertised for a very short time in ranching publications in Omaha and Chicago. This gave the Indians little or no time to protest or to organize their own cattle companies to compete with the established ranching outfits, who seemed to have the inside track.

The committee hearing was disrupted by an urgent telegram from three Sioux chiefs protesting the leases. "Four hundred families residing within the boundary of proposed lease oppose leasing to syndicate. Indians on reservation unanimously protest. Our farms will be overrun and trampled upon. Our efforts at home building and farming will be wasted. We ask you to investigate. Indians desire personal hearing. We are full-blooded chiefs." The telegram was signed Thunder Hawk, Walking Shooter, and Weasel Bear.[7]

The commissioner of Indian affairs, clearly shocked by the telegram, disavowed its significance and denied knowledge of the information contained in a subsequent letter signed by a Bishop W. H. Hare stating that the tribe had voted on a different lease arrangement than the one supported by the commissioner.

Despite overwhelming evidence that some agents were not acting on behalf of the Indians when they drew up the leases, Commissioner Jones shed new light on the problem by explaining that these Indian lands were already being overrun by thousands of head of cattle. The cattle were being illegally grazed by American ranchers, who used mixed-blood Indians to claim the cattle as their own, when in fact they belonged to outsiders. Jones argued that the leasing arrangement was simply an attempt to bring some control and intelligent management to chaos and rampant exploitation. "For many years the reservation has been practically stocked, mainly by outside cattle, the owners paying nothing whatever for their grazing privileges and being responsible to no one for any damage or injury they may inflict on the individual Indians," he said.[8]

This logical explanation left unanswered, however, the question of how the Indians would benefit. This apparently simple approach also left unanswered the question of who was actually responsible for leases on Indian lands: the tribes or the U.S. government? The Indian agents, Jones said, had based their actions on a legal opinion issued by the U.S. attorney's office on a ten-year-old law that stated that Indians could lease lands to which they had legal title. This narrow authority to lease had been extended to include all Indian land without a judicial interpretation or congressional approval. Jones said that leases had been issued for the past ten years on Indian lands throughout the United States, including a mineral lease on 640 acres of land on the Navajo Reservation issued to the Florence Mining Company.[9] (This is the first mention of a lease on Navajo lands, but the details have been lost to history.)

Ultimately the answer to the question of leasing rights on Indian lands was not based on morality or legality but on money and expediency. At the time of the hearings, Indians were in control of 50 million acres in the West. Politicians balked at this. It meant that unless they prevented the Indians from leasing these lands in their own right, the original Americans would reap the huge profits that could be made by exploiting the resources of the West. After all, the politicians argued, the West had not been "conquered" for the benefit of the Indians who had lived on it for thousands of years, but for the benefit of the new Americans who were convinced they were fulfilling their manifest destiny. The bill to open the Uintah Reservation to grazing and mineral leases was ultimately shelved, and the resolution of the issue was left for another day, effectively killing the Indians' chances to earn money from the lands they had been allotted.

The sentiment of the day was summarized by Senator Platt, the chairman of the Senate Committee on Indian Affairs: "This is a grave question. It involves the prosperity of almost the entire West.... In the carving out of these reservations, they include vast mineral regions in all those States.... I want every Senator to think of the responsibility to be assumed in throwing away half of the mineral lands of the United States....If it were intended by Congress that fifty million acres of land, which is three-quarters of the mineral lands of the country, were to be disposed of in that way, it would not have gotten a vote."[10] Once again Native Americans were effectively disenfranchised from their land, the only birthright they had, and condemned to a status as secondary citizens in their own country.

With the question of the tribes' right to lease their lands temporarily left hanging by congressional inaction, mining and mineral exploration continued. A movement to nationalize mineral lands received new impetus in 1910 from the recently created U.S. Bureau of Mines. Bureau engineers reported to Congress that the radium-containing ores found in America most likely were among the richest deposits of radium to be found in the world, but the ore was going to Europe because American miners lacked the technology to assay and refine it efficiently. It was being refined in Europe and used to cure cancer. Meanwhile, radium was virtually unavailable in the United States, and Americans were suffering. Bureau of Mines officials suggested that carnotite deposits should be nationalized to stem the flow of this valuable mineral to Europe.

When the bureau's report was released in 1913, the price of carnotite jumped 33 percent.[11] Dr. Robert Kelly, who was impressed with the medical possibilities of radium, joined with James Douglas, a chief engineer with Phelps, Dodge and Company, to propose a National Radium Institute that would be a business-government partnership to produce radium.[12]

In January 1914, Rep. Martin D. Foster of Illinois introduced House Joint Resolution 185 calling for the nationalization of carnotite, pitchblende, or other radium-bearing ores. Hearings began, but when it looked as if the resolution might pass, private mining companies in the West, specifically Standard Chemical, mustered massive opposition as they saw big profits being ripped from their hands. The effort to put the government in control of the exploitation of radium and carnotite died in Congress. This attempt to secure the uranium deposits on the Colorado Plateau would quietly succeed nearly thirty years later when the United States controlled all available uranium supplies to build the first atomic bomb.

Although nationalization efforts failed, the National Radium Institute was still alive, and its backers were determined to prove that government scientists were not bumbling incompetents but could lead the way in radium production and perhaps beat the market prices of private industry—all without any government subsidies. By the end of 1915, about seven grams of radium had been extracted at a cost of $37,599 per gram, which was less than a third of the world market price. The grams were divided between New York General Memorial Hospital and the Kelly Cancer Clinic.[13]

While the sponsors of the National Radium Institute proved their point, private industries such as Standard Chemical were not idle. The Flannery brothers found a new market for their radium with the outbreak of World War I and the discovery that a luminous paint could be created by mixing radium with phosphorescent materials. This paint was highly useful for nighttime warfare. Watch faces, machinery dials, even gun sights could be lighted.[14] World War I also created a huge demand for vanadium, which was being produced in large volumes by Vanadium Corporation of America (VCA), which was formed out of the Flannery brothers' Peruvian company. VCA became a major player in the vanadium market through World War II and was also a major supplier of uranium.

On June 30, 1919, the legality of leasing Indian lands was dramatically changed when Congress made its annual appropriation to the Bureau of Indian Affairs. This multimillion-dollar appropriation was a composite of all expenses relating to the American Indian. It included $10,000 for surveying Indian lands. It provided for irrigation projects ranging from the Klamath Reservation in Oregon to projects in Teec Nos Pos on the Navajo Reservation in New Mexico and Moencopi Wash on the Hopi Reservation near Tuba City, Arizona. It included $100,000 "for the suppression of the traffic in intoxicating liquors among Indians." It included $1.75 million to support Indian schools and school transportation.

Tacked onto the end of the appropriations bill was authorization for the secretary of the interior to open up Indian lands to prospecting and mining. The bill opened all Indian lands in Arizona, California, Idaho, Montana, Nevada, New Mexico, Oregon, Washington, and Wyoming that were "heretofore withdrawn from entry under the mining laws for the purpose of mining for deposits of gold, silver, copper and other valuable metalliferous minerals."[15]

The provisions of the laws were generous for the prospectors. Instead of being required to first request permission from the secretary of the interior to enter Indian lands for prospecting, anyone who had a pick, shovel, burro, and later a Geiger counter could enter Indian lands. Once they found a mineral deposit, they had to file a claim with the secretary of the interior; the claim was valid for a year. The department would then issue a lease that was good for twenty years and that could be renewed for up to ten years.

The leaseholder had to pay a rent of $1 per year for a tract of up to forty acres for a mining camp or a milling, smelting, or refining operation. The leaseholder was required to put in at least $100 per year of "development work," which could be either physical improvements or simply labor. For this work, just 5 percent of the net value of the mineral as it came out of the mine was to be paid to the government. The net value was the sale price of the ore after mining expenses had been paid. The government would place the royalties into funds that would annually flow to the tribe on whose lands the minerals were found.

The Indians were left to sit and watch their lands be exploited. They were given no control over who could mine, how it was done, or when and where it would be done. Reclamation was not mentioned. This basic arrangement for the leasing of Indian lands and payment of royalties remains in place today, although tribes have been given larger royalties and rents as well as other benefits, such as required employment of tribal members in mining operations.

In addition to opening Indian lands to mineral exploration and development, a brief section in the law also prohibited the president from creating new Indian reservations or expanding existing reservations by executive order. That right was suddenly reserved for Congress only. This meant, in effect, that there would be no new Indian lands unless Congress said so, which effectively placed the future of these lands in the hands of special interests and the congressmen they controlled.

With Indian lands now open to mineral exploration and the additional possibilities of finding oil and gas on the Navajo Reservation, the once-isolated lands were suddenly alive with prospectors. One of these was John Wade, who with his brother Jim five years earlier had opened up a trading post in the Navajo community known as Sweetwater. The Wades had been miners and prospectors in Colorado and Utah and knew of carnotite finds in the area. Since the trading business was slow, John Wade had spent his time wandering the Carrizo Mountains just to the east of Sweetwater and in 1918 had found many outcroppings of carnotite.

When Congress opened Indian lands to prospecting, Wade wasted no time and by 1920, under the names of the Carriso Uranium Company and the Radium Ores Company, had filed more than forty claims in the foothills of the Carrizos close to the Arizona and New Mexico border.[16] Production records are sketchy, but Wade must have been active because he paid a year's rental fees on 177.45 acres in May 1922. Sacks of ore were reported stacked at the Beclabito trading post, which was the closest post to the mine, albeit over very rugged terrain. These claims filed by Wade are the first known uranium claims and mines on Indian lands in the Southwest, and it was to these same claims that Wade and others, including some major companies like VCA and even the Atomic Energy Commission (AEC) would return twenty-five years later in search of uranium for atomic weapons.

In addition to Wade's company, which consisted of about a half-dozen mine groups, the Navajo Mining Company leased 80 acres, mostly in Arizona about two miles west of the trading post, and about fifteen hundred pounds of ore were shipped in 1921 from those claims. The total value of the ore was listed at $68.33 and of that amount, the Navajo tribe was to get 5 percent, or $3.81.[17]

As the worldwide demand for radium slowly increased and the demand for American vanadium grew despite the near stranglehold that VCA had on the

THE SWEETWATER TRADING POST OWNED BY JIM AND JOHN WADE AS IT LOOKS TODAY. JOHN WADE USED THE TRADING POST AS A BASE FROM WHICH HE DISCOVERED VANADIUM AND URANIUM DEPOSITS AROUND 1920.
PHOTO: PETER EICHSTAEDT

world market with its Peruvian mine, the future for carnotite ores must have seemed very bright to people like Wade and others who had found high-grade deposits on the Navajo Reservation. As quickly as this hope was ignited, however, it was dashed by a revelation by a Belgian firm. The Union Minère du Haut Katanga, headquartered in Oolen, Belgium, announced that it was selling radium for $70,000 per gram, about half of the going rate.

In 1913, the company had discovered what for many years was the richest pitchblende in the world, containing about 70 percent uranium. It came from the Shinkolobwe mines in the mountainous jungles of the southern Belgian Congo, close to the geographical heart of Africa. Amazingly, this find had been kept secret even though production was delayed for years during World War I. The ore was hauled by train out of the jungle on a railroad built from the mine to the west coast of Africa and then shipped to Belgium, where it was processed in a state-of-the-art facility.

Even the richest ores in North America contained only 2 percent uranium, which meant that it cost much more to obtain the same amount of radium. By 1923, the Flannery brothers' Standard Chemical closed its radium operations on the Colorado Plateau as the bottom fell out of the carnotite market.[18] The dreams of Wade and others who saw riches to be gleaned from the Navajo Reservation evaporated like mist in the harsh light of world competition. It was the end of carnotite mining on the Navajo Reservation and much of the Colorado Plateau until the exigencies of war revived it two decades later.

3 Secrets of the Earth

When the uranium and vanadium mines closed after 1923, life returned to normal for the Navajos living in Red Valley, a wide and dramatic expanse of land between the Carrizo Mountains in the north and the Lukachukai Mountains to the south. The wooden stakes that marked the 600-foot by 1,500-foot claims were knocked over by grazing sheep and the winds and rain that pelted the high desert valleys and mesas. During the fifteen years between 1923 and 1938, life was peaceful here. Meanwhile, in mines on other continents and in boardrooms of steel companies in the East, deals were being made that would greatly affect the lives of thousands of Navajos in the decades to come.

In 1926, the owners of U.S. Vanadium Company in Rifle, Colorado, sold a controlling interest to the Union Carbide and Carbon Corporation. The company's name was changed to the United States Vanadium Corporation (USVC), and it became a subsidiary of Union Carbide, a major supplier of ores that went into steel. It would eventually become a major miner and processor of uranium on the reservation. Its main competitor at the time was the Vanadium Corporation of America, whose deposits in Peru were dwindling rapidly. By 1928 these deposits were all but played out.

With the pending closure of the Peruvian mine, focus shifted to American sources of vanadium. In 1928, USVC moved quickly to buy up all the old carnotite leases that Standard Chemical held on the Colorado Plateau. This situation resulted in a strange but beneficial alliance between these two competitors. As a result, both companies were able to ride out the Great Depression of the 1930s. From 1933 to 1942, Union Carbide's USVC supplied most of the world's vanadium from mines on the Colorado Plateau that were not on Indian lands, and VCA sold it.[1]

In 1929, the New York Stock Market crashed, but no one heard the fall on the Navajo Reservation. In fact, on the Colorado Plateau, where just getting by was a way of life, the country's drastic economic downturn was barely noticed. There was so little activity in terms of prospecting and mining on the reservation at the time that in March 1936 the secretary of the interior closed the reservation to prospecting and the filing of claims.

Just two years later, however, Indian lands were once again opened to mining but with greatly revised procedures. These were prompted partly by the difficulty the Navajo Nation and government officials had in collecting the rents and royalties due for mines on Indian lands. On May 11, 1938, Congress adopted an act that gave tribal councils the authority to enter into lease agreements on reservation lands. The leases, however, had to be approved by the secretary of the interior. The new mining laws specified annual rents that increased every year and a base royalty payment of 10 percent of the value of the ore at the mouth of the mine. This amounted to a doubling of the previous payments. Again, bonds were required, and the total amount of land under lease for any one company or individual was limited to 960 acres. Leases were limited to ten years but could be extended if the mines were producing.

These procedures were later changed even further. Because numerous individuals requested leases on the same parcels of property, on April 9, 1941, the Navajo Tribal Council asked the secretary of the interior to allow leases to be put up for bid. Under this procedure, if an individual or company found a promising mineral location, they went to the tribal council and asked that the site be put up for bid. The lease went to the highest bidder. The initial leases could be written for large sections of land that were reduced in size after the specific mines were located.[2]

Not surprisingly, the revision to the mining laws that governed the exploration and development of vanadium and uranium on the Navajo Reservation coincided with an upswing in the demand for vanadium. The clouds of war were gathering in the skies of Europe and Asia in the late 1930s. Orders for ferrovanadium and vanadium steel from European nations and Japan increased dramatically as these countries armed themselves for the coming conflict. With this improved market, Vanadium Corporation of America in 1939 refurbished its vanadium processing mill in Naturita, Colorado, and began looking for some new, unclaimed sources on the Navajo Reservation.[3]

With the expansion of Nazi Germany under way, and after the invasions of Czechoslovakia and Poland, the United States began its own buildup of armaments, creating an unprecedented demand for vanadium. Soon the nation's ability to supply this critical war material for the Allies as well as its own needs was being stretched to the limit. In July 1940, the export of vanadium was placed under government control, and a year later vanadium suppliers first had to fill U.S. Department of Defense orders. Only leftovers were exported.

After the United States officially entered the war in December 1941, the restrictions on vanadium were tightened even further. The War Production Board, charged with ensuring the production of necessary war materials, organized the Metals Reserve Company in May 1942. This company signed an

FIRST MEETING OF THE WAR PRODUCTION BOARD IN WASHINGTON, D.C., ON JANUARY 20, 1942.
PHOTO: AP/WIDE WORLD PHOTOS

LEFT TO RIGHT: MARY EDWARDS, LUKE YAZZIE, ADAKAI, AND DENNY VILES OF THE VANADIUM CORPORATION OF AMERICA IN CANE VALLEY, ARIZONA, ON AUGUST 24, 1943.
PHOTO: PAGE EDWARDS

agreement with the USVC to act as its buying agent and ensure that an adequate amount of vanadium was supplied from the Colorado Plateau during the war. Although this agreement put USVC in a special and advantageous position to dominate the entire mining operations on the Colorado Plateau, it did not stop VCA officials from moving ahead with their own plans to take advantage of the huge vanadium market.

Because of previous mining experience over twenty years, John Wade as well as officials with the Vanadium Corporation of America knew that there were many unexplored vanadium deposits on the Navajo Reservation. Both private and government agencies appealed to the Navajo Nation to endorse an aggressive posture in the development of the mineral resources on their lands. In response, the Navajo Tribal Council approved a resolution supporting the development of the tribe's natural resources, with the provision that Navajos be employed to the greatest extent possible and that the tribal grazing land be disturbed as little as possible.

There was renewed interest in many "rediscovered" carnotite claims such as one ten miles south of Monument Valley on the Navajo Reservation. This site had been discovered by John Wetherill at the turn of the century but had gone unclaimed. The claim was ultimately located and surveyed by the U.S. Geological Survey (USGS). On October 1, 1942, a twenty-acre parcel of land at that location, known as the Combination Lode claim, was put up for bid. VCA was the highest bidder. It renamed the site Monument No. 1 and began work. The ore from that mine and other VCA holdings was sent to a vanadium processing plant owned and operated by VCA at Monticello, Utah, which was funded by the U.S. government's Defense Plant Corporation as part of the national war effort.[4]

The mining activity in the Monument Valley area did not go unnoticed by the hundreds of Navajos who lived in the area and maintained a marginal existence by raising sheep and selling wool rugs and blankets and an occasional concho belt to passing tourists at Goulding's trading post. Harry and Leone "Mike" Goulding had come to Monument Valley in 1924 and built a small, stone trading post and house into which they moved in 1926. The couple operated the store there for more than forty years, catering to film crews who used Monument Valley as a setting for many famous westerns, and to the miners and prospectors who combed the valley.

One of the Navajos who frequented Goulding's trading post was a young man named Luke Yazzie. He and his family lived a traditional Navajo life about ten miles away in an area known as Cane Valley. He heard that white men were looking for a certain kind of rock, and he was curious about it. When it was described to him, Luke knew where a lot of it could be found. He brought a sample of the gray rock to Harry Goulding, who turned it over for testing to Denny Viles, the production manager and vice president of VCA, who was in the area.

The results of the tests were astounding. It was one of the richest chunks of vanadium-uranium ore found to date anywhere in the four-state area. The rock was so rich that this mine, which would become known as Monument No. 2, would buoy the fortunes of VCA for the next twenty years. Word spread quickly, and soon the lease was put up for bid by the Department of the Interior's Office of Indian Affairs, which handled the advertising and bidding process for mining leases. VCA officials were intent on obtaining this lease because they knew what it could mean for the company. The lease was officially advertised on July 21, 1943, but when bids were opened on August 3, VCA was the only bidder. According to at least one disgruntled would-be bidder who was never notified of the lease offering, VCA was accused of bribing key individuals at the Navajo tribal headquarters in Window Rock, Arizona, so that the date of the bidding was not widely known.[5]

Fifty years later, Luke Yazzie sat beside his lifelong companion, June Yazzie, on a rug-covered couch in his modest frame house and recalled the days when he first brought white men to his backyard. Yazzie is a slight but fit man with a vivid memory and quick sense of humor that is evident in his eyes. Daily, he had climbed on his horse to check his flock of sheep. These graze in the sparse grass and brush that surrounded a once present pile of uranium mill tailings that were just a couple hundred yards from his house. Tailings consist of the gray, rough sand that was left over when the ore was converted into uranium oxide. Yazzie was about twenty-eight years old in October 1943 when he went to work at the VCA vanadium-uranium operation that was Monument No. 2. He was taught how to drill and blast, the typical method used for digging the vanadium-uranium ores.

As word spread that work was available in the area, the broad, long, rocky valley began to fill up with people, Yazzie recalled. Suddenly several hundred people were camped out in the valley. The scraggly piñons and junipers that tenaciously clung to life, sinking roots deep between the rocky cracks, were quickly cut for firewood. So many Navajos came looking for work that interpreters were hired to work on the site and a store was opened. Yazzie said he and others in the area were eager workers because they had been promised a share, albeit small, of the royalties. Those royalties never came.

"The workers were told they would get a portion of the profits," Yazzie said, speaking in Navajo as his daughter and daughter-in-law translated. With disgust in his voice, he said, "while we were working, Mr. Goulding got rich and moved to Phoenix." Yazzie said he periodically returned to Goulding's trading post and asked when he would be getting his royalties. "When he went to Goulding's he was told 'your money is coming,'" explained Violet Yazzie,

NAVAJO MINER LUKE YAZZIE SITS BESIDE HIS WIFE, JUNE, IN THEIR HOME IN CANE VALLEY, ARIZONA, IN FEBRUARY 1993.
PHOTO: PETER EICHSTAEDT

his daughter-in-law. "He never knew what was going on. He [Goulding] used to feed him [Yazzie] and stuff and act like [he] liked him." His nephew, Ben Stanley, said in a later interview that Yazzie's father warned Luke not to show the geologists where the uranium rocks were. "He said, 'Never take them to the white man. If you do, you'll get nothing out of it.' Luke took it to Goulding. He got a cigar for it."

While Yazzie and other Navajos waited for royalties on the vanadium-uranium ore, Goulding built a hotel and restaurant to accommodate the mushrooming number of visitors to Monument Valley. Today the approach to the magnificent valley is marked by huge billboards that offer a stay at Goulding's as a taste of life in the Old West.

Yazzie was not the only one wondering where his royalties went. Some miners who had leases adjacent to VCA's Monument No. 2 became suspicious when their obviously rich ore was being downgraded by VCA ore buyers. Through a friend they discovered that VCA officials were keeping two sets of books on the value of the ore. The company owed the Navajo Nation 10 percent of the value of the ore taken from the mine. The lower the value of the ore, the less royalty the company owed. In addition, VCA identified much of the

Monument No. 2 ore that it sold to the Metals Reserve Company as coming from sources outside the reservation in order to reduce the amount recorded as coming from this rich mine—and with it, the royalties that they owed the Navajos.[6]

The rapid development of vanadium mining in the Four Corners area caused by the war produced more vanadium in two years than anyone could imagine being needed in the foreseeable future, especially since the war seemed to be nearing an end, and so the Metals Reserve program was terminated on February 29, 1944, ending one of the biggest mining booms in the West. However, unknown to most of the rest of the world, two years earlier an entirely new mining boom had secretly begun that would dwarf the vanadium rush.

* * *

In 1896, when Becquerel and the Curies discovered the special properties of radioactive minerals, they could not have foreseen the improbable yet cataclysmic chain of events that would follow. The discovery of radioactivity itself indicated the natural breakdown of elements and the release of vast amounts of energy. When Albert Einstein formulated his theory of the conservation of energy, the implications were staggering. If the energy contained in even a tiny amount of matter could be tapped, it could provide mankind with an almost unlimited amount of energy to run machinery, heat homes, propel transportation. The possibilities were endless. On the dark side, it eventually became clear that such enormous amounts of energy could also result in the creation of awesome weapons of destruction.

The steps that followed the Curies' discovery were slow but steady. In 1919, British scientist Ernest Rutherford succeeded in bombarding nitrogen with radioactive particles and broke it into its component parts: oxygen and hydrogen. It was then discovered that of the component parts of the atom—the electron, the proton, and the neutron—the neutron was an uncharged particle and therefore, if it could be isolated, would be an ideal projectile to break up other atoms to release the energy they contained. By 1939, a number of scientists in the United States and Europe were deeply involved in fission research—attempts to break up the atom. They discovered that each time an atom was split it released energy and broke into smaller component parts, releasing a host of neutrons which went on to break up other uranium atoms in a continuous process.

Fortunately, at this time most of the key scientists working in this field had fled Germany or Europe as it was overrun by Nazis. These included some of the most brilliant minds of the century: Enrico Fermi, Niels Bohr, Lise Meitner, Otto Frisch, Edward Teller, and Leo Szilard—many of whom eventually were involved in making the first atomic bomb. These scientists, some in the United States and some in Europe, had conducted their own experiments that proved nuclear fission was possible.

LEO SZILARD, UNIVERSITY OF CHICAGO PHYSICIST, APPEARS BEFORE A CONGRESSIONAL COMMITTEE ON DECEMBER 10, 1945.
PHOTO: AP/WIDE WORLD PHOTOS

EDWARD TELLER, UNIVERSITY OF CHICAGO PHYSICIST, IN HIS OFFICE ON FEBRUARY 17, 1947.
PHOTO: AP/WIDE WORLD PHOTOS

One of the best fissionable products was a cousin of uranium. Normal uranium has an atomic weight of 238 (which indicates the number of protons it has), making it one of the "heaviest" elements in the world. In any amount of uranium, there is a small amount, less than 1 percent, of "light uranium," which has an atomic weight of 235. It is this light uranium that is considerably more unstable and therefore more fissionable than regular uranium.

When Germany overran Czechoslovakia in 1939, it captured the richest source of uranium in Europe at Joachimsthal. This little-known fact was generally overlooked by most of the world, which viewed with horror the political, not the scientific, implications of what was happening. It was not overlooked, however, by a few in the scientific community. Aware that German physicists were working on fission and that Germany might produce an atomic bomb, a small group of scientists, including Edward Teller, Leo Szilard, and Eugene Wigner, knew that the potential of atomic fission had to be brought to the attention of the government. However, it took persistent attempts by these scientists and others to get President Roosevelt to pay attention to the need for a U.S. program. The day before Pearl Harbor, a National Academy of Sciences committee headed by Vannevar Bush finally began implementing a program for military research on uranium, and in January 1942 Roosevelt approved the program to build the bomb—the Manhattan Project.[7]

An adequate supply of uranium was crucial to the project. The available sources of uranium in the world were the Colorado Plateau, Canadian mines, and the Shinkolobwe mine in the Belgian Congo. Members of the Manhattan Project immediately began combing the globe to secure all the sources of uranium that they could. By a stroke of luck, it was discovered that about 1,200 tons of high-grade (65 percent) uranium ore from the Shinkolobwe mine in the Belgian Congo were already in the United States.[8] This ore provided uranium for the initial experiments, such as the one conducted by Enrico Fermi and his group at Stagg Field, as well as trial efforts in refining uranium to obtain the highly fissionable uranium-235. Because the ore was so rich compared with that found on the North American continent, the United States quickly obtained options on the remaining stockpiles in Africa.

Attention focused on the Colorado Plateau and the Navajo Reservation, where vanadium mining was still in full swing. Geologists, miners, and atomic scientists knew the carnotite deposits also contained significant amounts of uranium. This suddenly valuable uranium was already stockpiled in the "waste" heaps at the four vanadium processing mills that had recently been built or renovated. (Until 1942 and the start of work on the bomb, uranium had been mined principally as a glazing element or obtained as a by-product of radium mining.) These mills included VCA's plants in Monticello, Utah, and Naturita, Colorado, and USVC's plants in Durango and Uravan, Colorado.

To get a more precise idea of the uranium available in the West, in late 1942 scientists on the Manhattan Project conducted a survey of vanadium deposits, mines, and milling stockpiles.[9] Among the mines studied were those in the Carrizo Mountains of the Navajo Reservation. The survey inadvertently revealed that the mining methods on the reservation were primitive at best. The work was being done as mining had been done for centuries, with pick, shovel, and wheelbarrow. In some instances bulldozers were used to push away dirt and rock that covered the uranium ore. Tunnels had been carved into the sides of exposed cliffs that contained the ore. By the new requirements of the leasing procedures, Navajo laborers were used to varying degrees in all mining operations.[10]

In January 1943, the Metals Reserve Company, which had contracted with both VCA and USVC to produce vanadium, developed agreements to sell uranium-vanadium sludges to the Manhattan Project.

WILLIAM CHENOWETH, RETIRED FEDERAL GEOLOGIST, LEFT, WITH PERRY CHARLEY, NAVAJO MINE RECLAMATION MANAGER, AT THE SYRACUSE MINE, RED VALLEY, ARIZONA, IN MAY 1993.
PHOTO: MURRAE HAYNES

In the summer and fall of 1943, three separate parties were sent into Red Valley to begin detailed exploration of uranium resources to find out which of the known deposits stood the best chance of producing large amounts of uranium. As it turned out, John Wade's Syracuse mine was found "to have the best prospective value of anything seen on the Navajo Reservation," according to one of the reports finished in 1944.[11] As a result of much of this exploration, the Union Mines Development Corporation purchased a 960-acre lease that was held jointly by Wade and the USVC, and which included many of the mines in the Red Valley area.[12]

By late 1944, about 4,000 tons of ore from the Belgian Congo, about 1,000 tons of ore from Canada, and about 1,000 tons of ore from the Colorado Plateau, including the Carrizo Mountains, had been refined, concentrated, separated, and converted into uranium metal needed to produce the atomic bomb.

4 The Uranium Boom and the Cold War

On August 1, 1946, almost a year to the day after the first atomic bomb was dropped on Japan, President Harry S. Truman signed into law the Atomic Energy Act of 1946 which created the Atomic Energy Commission, a civilian board (the military had wanted sole possession) that would oversee the development of atomic energy for military applications as well as for more peaceful uses like electricity. This act had a tremendous impact on the Navajo Nation and other tribes in the Southwest. It gave the government a broad mandate not only to explore all lands in the United States for the existence of uranium but also to condemn and obtain the land to secure the uranium, with or without the owner's consent. Primarily, however, it authorized the government to buy all the uranium it could find and gave it control over the extent to which uranium would be mined or not mined in the United States.

By the time the government announced a uranium purchasing program in April 1948, the ore was already being bought from the mines on the Navajo Reservation. Owing to the secrecy of the Manhattan Project, neither independent miners nor the Navajo Nation had been informed between 1943 and 1945 that uranium was being extracted from the vanadium-bearing ores. Word eventually leaked out that VCA and USVC, which owned the extraction mills, were being paid by the government for uranium as well as vanadium. It wasn't until April 14, 1944, that VCA agreed to pay the Navajo Nation for the uranium content of the ores as well as the vanadium.

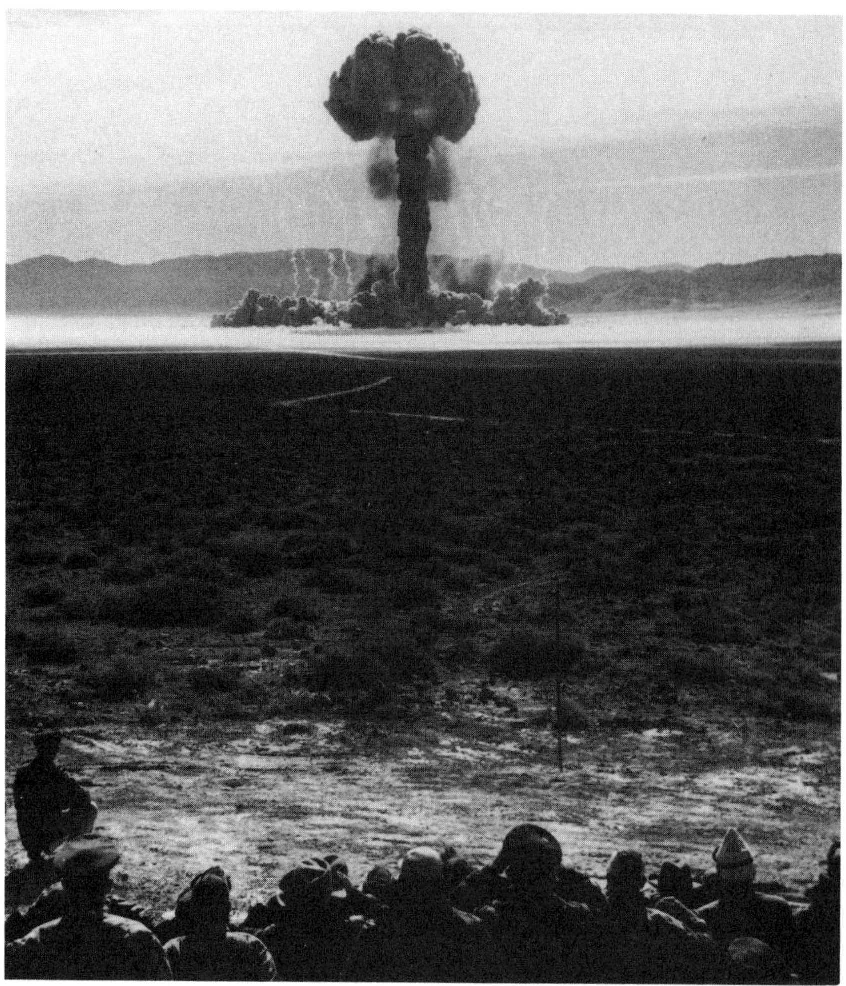

MILITARY OBSERVERS AND U.S. CONGRESSMEN WATCH AN ATOMIC EXPLOSION NORTH OF LAS VEGAS, NEVADA, ON MAY 25, 1953.
PHOTO: DEPARTMENT OF DEFENSE VIA AP/WIDE WORLD PHOTOS

The first royalty payment to the Navajo Nation for uranium came, not from the government's purchasing program, but from uranium used to color glass. In February 1944, a group of Blanding, Utah, men doing business as Carroll and Shumway obtained a lease to mine on Oljato Mesa, a dramatic butte in Monument Valley. The mesa top was mined from June to December 1944, producing more than 100,000 pounds of ore that contained about 4,000 pounds of uranium and nearly 6,000 pounds of vanadium. The ore was dug by hand, using picks and shovels, put into wheelbarrows, and lowered over the side of the mesa in buckets. Eventually it was shipped to the Vitro Manufacturing Company in Canonsburg, Pennsylvania, for processing. Royalties being pegged at 10 percent, the Navajo Nation eventually received $606.13 for the uranium ore that was mined from the top of the mesa.[1] It seems a small sum compared with the costs in lives and environmental damage that another forty years of uranium mining on and around the Navajo Reservation would cause.

In 1947, the AEC signed uranium production contracts with companies like VCA and USVC. The former reopened many vanadium mines in the Carrizo Mountains and Monument Valley areas of the Navajo Reservation as part of an overall AEC plan announced in April 1948 to stimulate uranium mining on the Colorado Plateau. The drive to develop domestic sources of uranium was launched when it was learned that the rich reserves in the Belgian Congo, from which most of the early uranium had come, were nearly played out. There were other discoveries of uranium in the world, such as in Canada, South Africa, and eventually in Australia, but the reliability of these reserves and the available transportation were questionable. However, the AEC continued to purchase uranium from Canada and South Africa through 1967.

Details of the new AEC ore-buying program were provided in a series of public circulars first issued on April 11, 1948, that announced the government's price for uranium ore and uranium concentrates from the mills. The price for raw ore would be based on the uranium content. First the ore would be tested for uranium percentages at an ore-buying station, then weighed. Calculating the price of the ore was a complex arrangement that included premiums for higher grades, hauling allowances, and bonuses for new discoveries. The uranium ore on the Colorado Plateau was relatively low grade and averaged only 0.25 percent. This meant that about 5 pounds of uranium oxide were in each ton of raw ore. A price of $3.50 per pound of uranium was paid for the ore. Thus, a ton of 0.25 percent ore was worth $17.50. An incentive program offered a $10,000 bonus for the delivery of 20 tons of concentrated ore with at least 20 percent uranium oxide from any new, previously undiscovered location. The AEC would pay miners for hauling costs based on distance as well as the costs of new equipment needed to develop new mines.[2] With all possible bonuses, the final price for a ton of 0.25 percent ore could go as high as $38.25.

The AEC contracts with the mills also specified the price of concentrated uranium oxide, known as yellowcake, the product that came out of the uranium mills and which was further refined into uranium metal. The price of yellowcake started at $7.14 per pound in 1948, increased to a high of $12.51 per pound in 1955, and then dropped to $8.75 by 1960; it finished at $5.54 in 1970 when the uranium buying program ended. The new price of $7.14 per pound for uranium concentrate in 1948 was a big improvement over the 77¢ per pound paid through World War II.

The government's proposal to develop America's uranium reserves quickly took on militarylike proportions. Before it was all over, ore-buying stations were opened in Marysvale, White Canyon, and Moab, Utah; Shiprock and Grants, New Mexico; Globe and Tuba City, Arizona; Riverton and Crooks Gap, Wyoming; and at Edgemont, South Dakota. In addition, the AEC encouraged temporary ore-buying stations to be set up while mills were being built at Bluewater, New Mexico; Salt Lake City and Mexican Hat, Utah; the Shirley Basin area in Wyoming; and in Karnes County, Texas.[3]

The AEC's exploration program was enormous. As a result of surveys on the Navajo Reservation and in the Four Corners area, exploration spread in all directions. Working with the AEC, the U.S. Geological Survey assigned more than 100 geologists to help with the search for uranium. Between 1949 and 1954, about 500,000 acres of public lands were legally withdrawn from public use so that geologists could conduct drilling projects unimpeded. Government exploration included drilling, radiation measurements, airborne reconnaissance, and exploration camps.[4]

From 1948 to 1956, the USGS and the AEC spent millions of dollars drilling exploratory holes near possible uranium deposits. If promising core samples were found, individuals or companies could go about securing leases. More than 7 million pounds of uranium were found this way. From 1948 to 1958, the AEC and USGS studied about 7,500 reports of uranium finds in forty-two states. The AEC also took to the sky, in 1953 establishing its own air force of ten Piper Supercub fixed-wing aircraft that flew over 81,000 square miles of remote lands and chalked up 1,140 uranium discoveries in eleven western states from California to Arkansas. The AEC also spent $17 million building roads to important deposits and by 1960 had built or improved 1,253 miles of access roads.[5]

As the production of ore increased because of the valuable incentives the AEC offered, it quickly outpaced the capacity of the three existing mills in the West that produced yellowcake. These were the AEC mill in Monticello, Utah, the VCA mill in Naturita, Colorado, and the USVC mill in Rifle, Colorado. Additional contracts opened a pilot plant mill in White Canyon, Utah; and mills in Durango, Uravan, and Grand Junction, Colorado; Salt Lake City, Utah; and Bluewater, New Mexico.

On August 7, 1953, the AEC signed a contract with Kerr-McGee Oil Industries to build a new mill at Shiprock, New Mexico, on the Navajo Reservation. The mill would process the ore deposits that were being worked in the Lukachukai Mountains, forty miles west of Shiprock. Constructed on the banks of the San Juan River, the mill would operate until 1968. By 1960, there were twenty-seven uranium processing mills in operation, stretching from Washington to Texas.[6]

U.S. VANADIUM COMPANY MILL ON MAY 15, 1952. THE MILL PRODUCED VANADIUM AND URANIUM AT URAVAN, COLORADO.
PHOTO: AP/WIDE WORLD PHOTOS

Almost as quickly as the AEC program expanded, it contracted. In just one decade so many deposits of uranium had been found that there was enough ore to supply the military and domestic energy needs of the country far into the foreseeable future. In 1957 and 1958, the AEC shut off the rapid expansion of the industry by announcing that after April 1, 1962, it would no longer buy ore and would only purchase uranium concentrate from ores discovered prior to November 24, 1958, at a fixed price of $8.00 a pound.

* * *

The Navajo Nation early on was approached by the AEC for approval and cooperation in the development of the uranium reserves that geologists knew were scattered across the Indian lands. In the late 1940s and early 1950s, the Navajo Tribal Council approved a series of resolutions that encouraged the development of uranium ores and gave Navajos preferential treatment in this development. One such resolution unanimously approved on October 14, 1949, said the following:

> Whereas the Navajo Tribe owns and controls mining properties containing deposits of uranium-bearing ores including other ore, and,
>
> Whereas the United States government, as well as many nations throughout the world are vigorously seeking sources of uranium-bearing ores for development of atomic energy whether for weapons of war or for peaceful and beneficial purposes and,
>
> Whereas the present rules and procedures of the Navajo Tribal Council pertaining to the development and mining of uranium-bearing ores are not conducive to progress in and prospecting for and development of mining, with the result that these ores are not being supplied to the United States Government in the quantities desired, and consequently revenues in the form of rents and royalties are not accruing to the Tribe as should be the result if exploration of these minerals can be increased....[7]

The wording shows that the Navajo leadership was convinced that there was a lot of economic benefit to opening up their lands to uranium development.

Yet at the same time that the tribal officials were talking with the AEC, studies being conducted on the uranium mines on the tribe's lands were revealing highly dangerous working conditions. While tribal officials were asking for more development, the Navajo miners were being exposed to high levels of radon gas, radiation, and radioactive dust.

The Navajo tribal resolution went on to establish an advisory committee to the council that would set up rules, regulations, and procedures so that the interests of individual members of the tribe would be protected and individual Navajo initiatives in mining encouraged. The resolution set a 10 percent minimum royalty and said that individual Navajos who obtained leases could receive a portion of the royalties. It authorized the AEC and the USGS field representatives to explore and drill on Navajo lands as long as the information derived was turned over to the tribe.

The practical result of the resolution was achieved in that it brought many individual Navajos into uranium mining as entrepreneurs, not just as miners. However, it was difficult for most Navajos to benefit substantially because they lacked the equipment and expertise needed. A typical arrangement was for a mining company to obtain a mining permit from a Navajo in exchange for 2 percent royalties on production. The company would then hire the Navajo as a foreman and have him hire and supervise his family and friends as miners. After the richest and most profitable ore was mined, mining companies then would sell the used equipment, drop the assignment of the mining permit, and let the Navajo reap what little profits he could from the remaining low-grade ores. In some cases, in order to make some money in a managerial operation, the Navajo subcontractors paid their laborers well below minimum wage—as little as $1 a day—and cut corners by not ventilating the mines.

A TYPICAL URANIUM ORE BUYING STATION OWNED BY THE U.S. VANADIUM COMPANY IN THOMPSON, UTAH. THE ORE WAS TESTED AND MINERS PAID BASED ON THE URANIUM CONTENT OF THE ORE. PHOTO: AP/WIDE WORLD PHOTOS

The AEC's buying structure meant that a relatively rich ore deposit, such as a 0.5 percent grade, could obtain prices ranging from $35.00 to $79.50 per ton of ore, depending on bonuses and development allowances. If a rich ore deposit was played out and a Navajo entrepreneur was left with a permit to a mine with ore grades of only 0.15 percent uranium, the ore would only bring prices of $7.50 to $16.50 per ton.

The permit to VCA's abandoned Monument No. 1 mine, in southern Monument Valley, and one of its earliest vanadium mines, was acquired by Cecil Parrish, Jr., a Navajo who lived in the area. Parrish in turn assigned the permit to two non-Navajos—Charles Ashcroft, Sr., and J. L. Foutz, a trading post owner from Farmington, New Mexico. Parrish acquired a second permit adjacent to the first one, which also produced some profitable amounts of uranium. Although bonuses and extra royalties were to have been paid to the Navajos for their ownership of the permits, Parrish claimed during an interview that he did not receive money for his involvement in the permits or the mines but worked as a mine manager who hired and supervised many of his fellow Navajos. These men worked long shifts in the mines under crude and miserable conditions.

Some Navajos, such as Cato Sells of Farmington, were able to make money as developers. VCA resumed operations in 1947 at its Monument No. 2 mine, digging the same ore now for uranium instead of the vanadium. Cato Sells and several other Navajos had obtained permits for land surrounding the original VCA lease, and because of the value of the finds, VCA acquired the Indians' permits in July 1959 and expanded their own 1942 lease.[8]

In the eastern Carrizo Mountains near Red Valley, Cato Sells and another Navajo, Nakai Chee Begay, operated what was known as the Upper Canyon mine and the Red Wash mine. Mining in the area was stimulated when the AEC opened up its ore-buying station at Shiprock, in January 1952, providing a new and readily accessible market for the ore in the area.[9]

* * *

By early 1951, word spread on the Colorado Plateau that a new uranium deposit had been found in central-western New Mexico near Grants by a Navajo named Paddy Martinez. Martinez had spent a lifetime wandering the arid, rocky mesas with a flock of sheep. He learned that white men were looking in the area for a certain kind of rock. He asked about it and was shown what the rock looked like. Then he took geologists to what would become one of the largest uranium districts in the West.

Martinez led the geologists to a uranium-bearing ore called tyuyamunite at the base of Haystack Butte on property owned by the Santa Fe Pacific Railway Company, a land-owning subsidiary of the Atchison, Topeka and Santa Fe Railway Company. Martinez's find would lead to other equally dramatic discoveries in what would become known as the Grants mineral belt, a vast and rich deposit extending from the Navajo Reservation north of Gallup several hundred miles east to the east side of Mount Taylor.

The railway company immediately formed a mining subsidiary called the Haystack Mountain Development Company and began an extensive exploration program to determine the extent of the find and the possibilities of other deposits on railroad lands. Word spread quickly, and soon the Anaconda Copper Company as well as the AEC joined in the search, sending in a team of geologists. Their glowing report was made public in March 1952 and spurred further searches for uranium.

Deposits had been discovered in the nearby Poison Canyon area in 1951. In the spring of 1955, drilling by Louis Lothman led to additional large uranium discoveries in the Ambrosia Lake area north of Grants. These deposits were developed by the large companies of Kerr-McGee Oil Industries, which was already involved in the uranium industry in Shiprock, by Homestake Mining Company, and the Phillips Petroleum Company. The finds were so large that they led to the construction of four processing mills in the area.[10] In addition, the Anaconda Copper Company made a new discovery on lands of Laguna Pueblo, fifty miles west of Albuquerque, that would become the Jackpile mine, one of the largest-producing, open-pit uranium mines in the world.

Despite the resolutions passed by the Navajo Tribal Council in the late 1940s and the early 1950s to provide priorities for the Navajos in prospecting and mining uranium, William Chenoweth, an AEC geologist who worked in the area during the 1950s and 1960s, said the system was badly abused:

> Here'd be a family herding sheep in this area, having their hogan in this area, and maybe a white prospector could find an outcrop up near their place. And he'd need a mining permit.
>
> Well, if you found something with that prospecting permit, only a Navajo could apply for a mining permit. And you could use any Navajo in the world. So here'd be this family living here on this land for many, many years. But if you knew an English-speaking Navajo that was kind of bright, and knew the politics of Window Rock, you use[d] him to go get a permit. And if a mine was developed, your permittee, your Navajo partner, would get royalty (2%), but the people living there wouldn't get a thing. This was a very abused [system], this was very abused, especially in Cameron. And there were a lot of injustices done that way, but that was the way that politics work[ed].[11]

While the system was abused by people who understood how to manipulate it, there were others who used it to their advantage. Chenoweth recalled one such person in the Oakspring area, a woman named Despah Tutt, who was married to a Navajo man known as King Tutt. Hoping to reap some of the benefits of possible uranium finds near her house, she often insisted that AEC rigs continue drilling deeper even though geologists said it was useless.

It is no wonder that Navajos and other Native Americans today are resentful of the white man's intrusions and exploitation of the Indians. Chenoweth explained that it became a dark joke among the Navajos that each time white men arrived, they wanted something from the Indians. Until the AEC arrived on the northeastern section of the Navajo Reservation in the early 1950s, there had been few white men except for people like John Wetherill and the Wades, who were traders as well as prospectors, and who made friends with the Indians. Chenoweth said the story was told that each time a group of white men arrived, the Navajos asked: What are you looking for? In the 1920s the answer was oil; in the late 1930s and early 1940s it was vanadium; in the 1950s, 1960s, and 1970s it was uranium. In the 1990s it is spirituality.

While Navajos were given limited opportunities to reap the benefits of the uranium that was being taken off their land, they were given a variety of unskilled, minimum-wage jobs even with the AEC, Chenoweth recalled. "One other thing...that I remember working out there in the exploration camps that was...we hired local Navajos to...empty the trash, and haul water, and just general maintenance around there. And they paid them minimum wage...." The local Navajos were more than willing to work, Chenoweth recalled, but the opportunities were limited.[12]

As he worked for the AEC in various capacities, Chenoweth witnessed the development of uranium mining on the Navajo Reservation and the Colorado Plateau. He recalled that on the reservation, particularly in the Lukachukai Mountains, working conditions were primitive but generally followed practices at the time. Depending on the size of the mine, there were from four to fifty miners working in shifts. In the Carrizo Mountains, north of the Lukachukais, there were smaller mines with only four to fifteen men per shift and only about one shift per day. Pay was less than $2.00 per hour, depending on the duties, and with Kerr-McGee, the pay was generally about $1.50 to $1.70 per hour.

Millions were being made in the uranium industry in the 1950s, and in response to thousands of requests from people with dreams of entering the business, the U.S. Bureau of Mines issued a series of information circulars. These described, in the most concrete terms possible, how uranium mining was done. One circular looked at the operations of three mines owned by the Climax Uranium Company, including the Frank No. 1 mine, three miles south of Cove in the Lukachukai Mountains. The circular notes that the mine's crew was entirely Navajo. The mine opened in April 1951 and continued to produce through 1958. Frank No. 1 is a mine in the steep hillsides where the Mesa mines were found, and was named after Frank Nacheenbetah, a mining foreman who discovered the deposits in 1950 and to whom the permit and royalties were assigned. He turned over the operations to Climax in early 1951.

The Bureau of Mines circular describes the following conditions: "On the

A NAVAJO MINER EMPTIES FRACTURED URANIUM-BEARING SANDSTONE IN THE FRANK NO. 1 MINE NEAR RED VALLEY, ARIZONA, IN THE LATE 1950S.
PHOTO: U.S. BUREAU OF MINES

stopes [a steplike excavation] above track level, wheelbarrows are used to transport the ore to the haulage-level chutes. Nacheenbetah reported this method of transportation is well adapted to the small irregular undulating stopes. Hand loading has enabled the miners to keep the ore clean. Burro haulage and hand tramming are used to transport the end-dump 16-cubic-foot mine cars through the adits [tunnels]."[13]

The ore was loaded into trucks and hauled to the Climax uranium mill in Grand Junction, Colorado, for processing. No mention is made of the dangers of exposure to radiation as well as radon. Since none of the work was mechanized, the miners were fully exposed to hazardous conditions.

The circular went on to say:

> The Navajo Tribal Council stresses that Navajo workers be employed in the mines on the reservation. As a whole, the Navajos make good uranium miners and have a keen knack of knowing waste from ore. Their patience is undoubtedly a help where the irregular ore lenses must be broken and handled selectively.

ONE OF THE MESA MINES IN THE LUKACHUKAI MOUNTAINS SOUTH OF RED VALLEY, ARIZONA.
PHOTO: MURRAE HAYNES

> Seventeen men, working one shift a day, are employed at this 950-ton-a-month operation. The size of the crew has remained unchanged since production started in 1951. Besides the mine foreman, 13 men work underground and 4 on the surface. An average of nine muckers and trammers is needed to handle the rock broken by four drillers. In May 1958, miners were paid $1.91 an hour and muckers and trammers $1.55 an hour. As would be expected, the productivity of labor is lower at the Frank mine than at the other two [Climax] operations. In 1956 the average tons produced per manshift for all labor and supervision at the mine was 2.47, and the average of both ore and waste was 4.33 tons per manshift.[14]

Productivity was relatively low, the circular pointed out, not because of the abilities of the Navajo miners but because of the difficulty in obtaining high-grade uranium ore. The ore was in a much thinner layer and was very irregular, which meant that all of the work had to be done by hand. There was only minimal mechanization. As a result, production was lower than in mechanized mines, and the labor costs per ton were higher.

The Kerr-McGee mining operations in the Lukachukai Mountains employed from sixty-five to seventy Navajo miners supported by thirteen supervisors, all white men. These ranged from the superintendent of operations to the camp cook. The company built a so-called mining camp in Cove that consisted of prefabricated buildings in which the thirteen supervisors lived during the week. On weekends they commuted from Cove to their homes in Cortez and Farmington. The supervisors paid $2.00 a month to the company for the housing and 50¢ a meal. The Navajo miners, meanwhile, received no housing or subsidized meals and were left to make their own way between mine and home, often covered with mine dust.

The clash of cultures is apparent when the Kerr-McGee company describes its experience working with the Navajo miners:

> When Kerr-McGee first began mining on the reservation, the turnover of Navajo labor was very high, mainly because many of the men did not easily accept routine. Normal to their way of life, the Indians worked only long enough to satisfy their immediate needs. Attendance improved steadily and by mid-1958 Kerr-McGee had a good core of reliable men averaging three years of service with the company. Labor turnover still was approximately 20 percent every six months. Absenteeism without prior notification was at one time of major concern, but this also lessened. Although the Navajo are conscientious and hard-working miners, it is only within the last few years that they have appreciated the need for reporting daily to the job.[15]

Such differences were never more apparent than when it came to life-and-death situations. "The company has conducted regular biweekly first aid and safety meetings. At these meetings safety at home and on the highway is also stressed. One of the principal aims of these meetings is to acquaint the Navajo with the need and obligation to give first-aid treatment to a person who is injured or ill until professional aid is obtained. And ancient superstitious fear of the dead and the dying is fading, but is still strong enough to stop some of them from giving aid to an injured or unconscious person."[16]

A circular on the Kerr-McGee uranium mine operations in the Mesa mines in the Lukachukai Mountains describes how the leases, mining permits, and accompanying royalty payments were assigned to relatives of a Navajo, Dan Phillips. The relatives received royalty payments of at least 2 percent of the value of a ton of ore at the mine. The royalty payment increased from 2 percent up to 5 percent as the value of the ore increased from a low of $30.00 per ton to $80.00 per ton. Although the price of uranium oxide ranged from a low of about $7.14 per pound to a high of $12.51, it averaged about $8.00 per pound over the years. Depending on the quality, a ton of ore, at 2,000 pounds, would have about from 2 to 5 pounds of uranium in it. If each miner could produce about 2.5 tons per eight-hour shift, that would mean from $40.00 to $100.00 worth of uranium produced per day per miner. Out of that money the mine company would pay from $1.50 to $2.00 per hour or $12.00 to $16.00 per shift per miner for labor.

According to regulations, the Navajo Nation also received a graduated royalty beginning at 10 percent of the value of the ore. This increased to 20 percent as the value of the ore rose from a low of $10.00 to a theoretical $100.00 per ton. As a result, a concentration of only 0.15 percent for uranium ore made such mining operations a low-profit proposition. It was these marginal operations that were often turned over to Navajo mine operators after the larger companies realized that the ore grades were dropping, along with their profits. The grade of the ore was crucial to the profit margin. If the deposits contained vanadium, this ore, which remained valuable, contributed to mining profits.

Although some large mining operations had slim profit margins due to low-grade ores, profits were improved in many cases when companies owned

AN OFFICE SOLICITS URANIUM ORE BUSINESS ON MAIN STREET, GRAND JUNCTION, COLORADO, IN JUNE 1955.
PHOTO: AP/WIDE WORLD PHOTOS

other richer deposits or built uranium processing mills, as Kerr-McGee did in Shiprock. The government fixed the price of yellowcake produced by the mill, and after the construction costs were paid, the profit margin at the mills was much higher than at the mines.

Given their generally high profits, it is inexcusable that large mine operators resisted later efforts by the government to require ventilation in the mines and respirators for the miners. Since many small mine operators, typically the Navajos and independent miners, worked the low-grade deposits, it is understandable that they could not afford ventilation and other protection.

5 The Shadow of Death

IN ONE OF THE STORIES the Navajos tell about their origin, the *Dineh* (the people) emerged from the third world into the fourth and present world and were given a choice. They were told to choose between two yellow powders. One was yellow dust from the rocks, and the other was corn pollen. The *Dineh* chose corn pollen, and the gods nodded in assent. They also issued a warning. Having chosen the corn pollen, the Navajos were to leave the yellow dust in the ground. If it was ever removed, it would bring evil.[1]

* * *

By the time the Atomic Energy Act of 1946 was signed into law, the medical community and the government were very aware of the dangers of radiation. Doctors had known for years about the mysterious "mountain disease" that afflicted miners in Czechoslovakia. The disease had baffled doctors because otherwise healthy miners had died within six months of contracting the disease. The culprit, scientists and doctors eventually believed, was radon.

As uranium breaks down into other elements, the energy released has three different forms: alpha and beta particles, and gamma rays. Alpha particles are potent but can be stopped easily by such things as a sheet of paper or even human skin. However, this does not mean these little particles are harmless. Once alpha particles are taken into the human body, they lodge in tissues, bones, or organs, and steadily radiate and pelt surrounding cells. Beta particles are very

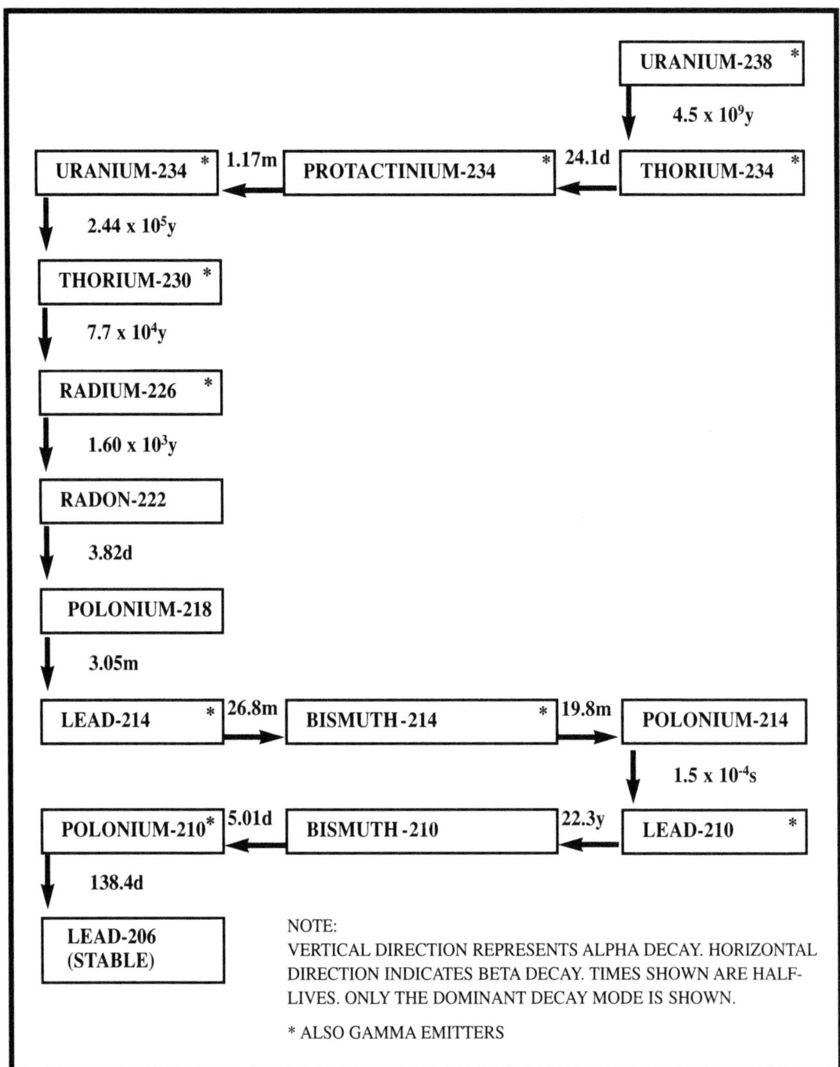

Radioactive Decay Chain of Uranium-238

Source: *Phase II - Title I Engineering Assessment of Inactive Uranium Mill Tailings*, U.S. Energy Research and Development Administration, March 31, 1977, prepared for U.S. Energy Research and Development Administration, Grand Junction, Colorado, by Ford, Bacon & Davis Utah Inc., Salt Lake City, Utah.

similar, but thicker, denser materials are needed to stop them, and they can burn skin. Once inhaled they can wreak havoc in the body. Gamma rays are highly penetrating rays that require about an inch of lead or a foot of concrete to be stopped. These rays pass right through the human body. The earth is bombarded constantly by gamma radiation from the cosmos, but intense doses from such earthbound elements as radium can be fatal.

As uranium moves through its natural decay process, it is transformed into a variety of different elements until it finally becomes a stable form of lead. Some of these steps in the uranium decay chain last only a few minutes or days, while other steps take thousands of years.

The time it takes for any amount of these elements to break down into the next element in the chain is called the half-life. For example, radium breaks down after 1,600 years into the gas radon, which in turn converts or breaks down with a half-life of 3.8 days into a solid particle of polonium, which breaks down in just 3 minutes to lead-214, which is radioactive and which breaks down in 26.8 minutes to bismuth-214, which in another 19 minutes becomes polonium-214, and so on. The elements that follow the rapid breakdown of radon have become euphemistically known as "radon daughters," and it is these elements that cause lung cancer.

Exposure to radon daughters was a risk for the uranium miners because of the radon that oozed out of the mine walls or, because it is soluble in water, out of the water in the mine. The normal decay process would release a steady stream of radon daughters into the air. These would become attached to dust particles or cluster together as molecules. Since much of the work in the mine was done with a pick and shovel, the miners' breathing rates were often accelerated, which meant that they took in sizable amounts of radon daughters. The particles would lodge in the miners' lungs or air passages, where alpha particles would bombard the tissues.

In the uranium mines, sensitive lung tissues were constantly subjected to small but steady doses of radiation. These small, steady doses have recently been found to be more likely to cause cancer than a single heavy dose of radiation. When penetrating radiation strikes genetic material, a cell is damaged. Normally, damaged cells die and are replaced by new cells; however, sometimes the injured cells survive but are incorrectly repaired. This leads to mutations and, sometimes, cancer. The earliest lung cancers found among the miners were what is known as the oat cell type, producing death within six months. Later work showed that other types of lung cancer were also related to mining.

While breathing radon daughters was one of the grave dangers of uranium mining, a second dangerous problem was the dust from the rocks. The normal mining technique at the time was to drill holes in the soft sandstone rocks, put dynamite in the holes which had the highest radiation levels, and blast the rocks. When the sandstone was exploded, it created thick dust that hung in the dead air. Usually the blasting was done at the end of the day, and the dust was allowed to settle overnight. When the miners returned the next morning, the broken rock was removed, dumped into an ore bin, loaded in trucks, and hauled to the mills.

For the first ten years, few of the underground mines had ventilation. Many operators of small, marginally producing mines never installed ventilation in the mines because it would have reduced their slim profits. Others, eager to remove as much uranium as possible per hour, regularly sent workers back into the mines within minutes after blasting. The prime component of sandstone is quartz or silica, a glasslike material. Repeated inhalation of silica dust can produce scarring and reduced lung capacity, a condition known as silicosis.

Timothy Hugh-Benally, who was also a miner, in February 1993 recalled the conditions of the Kerr-McGee mines in the Cove area of Red Valley. From his office, he spoke of those days:

> The working conditions were terrible. Inspectors looked at the vents [ventilation fans]. When they weren't inspected, they were left alone. Sometimes the machines [for ventilation] didn't work.... They told the miners to go in and get the ore shortly after the explosions when the smoke was thick and the timbers were not in place. There was always the danger of the ceiling coming down on them.
>
> The foremen were Anglos [white men] and rarely went into the mines. I complained about that. I said if we were represented by a union, we'd get our rights. I was fired again. I left it.... The people I worked with are now all dead from cancer or other causes.... A lot [of those living] have filed claims, and some are still waiting. It's kinda sad to see people come and file claims and their claims are held up for various reasons. We filed some test cases and saw how they are processed. We'll see what happens.
>
> The people who worked in the mines started dying in the 1950s. We tried traditional ceremonies to cure the husbands. We tried traditional remedies, and some tried the Native American Church. They gradually went down. They were usually heavyset men, and when they died they were skin and bone.

Dan N. Benally, age seventy-seven, a resident of Red Valley, worked in the Kerr-McGee mines in the Cove area. He speaks only Navajo and understands some English. He was injured severely in a rock fall and lost his sight in his left eye. Through a translator, he, too, recalled the working conditions in the mines while he took a break from a Red Valley Chapter House meeting in early 1993:

> He was never warned how it would affect him in the future. They used to eat underground and drink the water dripping from the walls.... When they did the blasting, they inhaled the smoke and dust. He fainted twice, and they had to drag him out.... He worked with thirty-eight people in Cove, and all of them died. One of his relatives passed away recently.
>
> He was never told or warned [of the dangers]. He was not told to wash his hands, and [they] were told to stay underground. He worked shifts from 7 A.M. to 8 P.M. with one hour off for lunch, and earned $1.15 per hour.... In the springtime he can still taste the gunpowder.

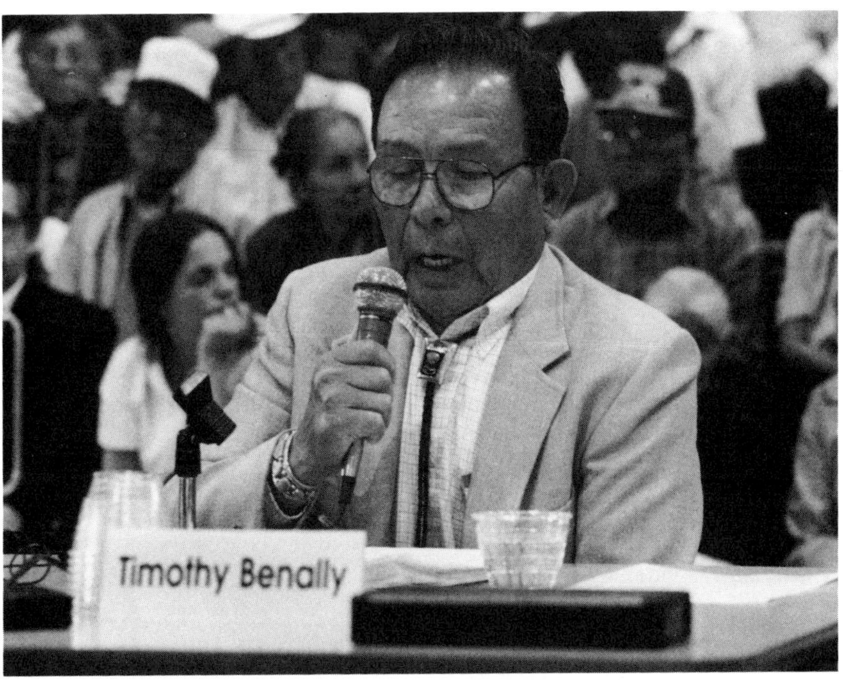

TIMOTHY HUGH-BENALLY, A FORMER NAVAJO MINER, TESTIFIES AT A 1993 CONGRESSIONAL HEARING.
PHOTO: MURRAE HAYNES

As early as November 1949 the U.S. Public Health Service (PHS) was worried about the working conditions in the uranium mines and possible health effects on miners. In response to questions about the safety of the mines from the interior department's Office of Indian Affairs, the PHS surveyed mines in three areas on the Navajo Reservation, all leased and operated by the Vanadium Corporation of America. The exploratory study was done by Henry Doyle, an engineer working for the Public Health Service. The PHS inspected several mines in Red Valley west of Shiprock known as the King Tutt mines (the East Reservation lease), VCA mines on the north end of the Carrizo Mountains (the West Reservation lease), and the Monument No. 2 mine in Cane Valley south of Mexican Hat.

Radiation measurements in the mines were recorded in milliroentgens, an indication of the amount of energy in a liter of air. Contemporary measurements are done in "millirems," which are a slightly larger quantity of energy. A milliroentgen is approximately 90 percent of a millirem. The recordings taken in the King Tutt mines showed from 0.5 to 1 milliroentgen per hour of both gamma and beta radiation. Measurements of 1.5 milliroentgens were taken at the mouth of the mines. Contemporary standards allow for 2 millirems per hour for miners but limit the total dose for seven days to 100 millirems; the annual allowable dose is set at 500 millirems. According to contemporary standards, the workers were under the allowable doses on an hourly and weekly basis, but annual exposure greatly exceeded the allowable doses. These measurements were simply external measurements, however, and did not include additional doses of radiation the miners received from both the radon daughters and the uranium inhaled in the dust.

Health officials were already worried that contact with skin and impregnation of the miners' clothing with uranium dust was a potential problem. "When the instrument was held several centimeters from the worker's clothing and skin, the radiation, considering both Beta and Gamma, was in the vicinity of 1.5 milliroentgens per hour."[2]

When measurements were taken at the VCA's Monument No. 2 mine, the radiation readings were shocking. Part of the mine at the time was open pit, which allowed radon gas and uranium dust to disperse. However, other parts of it were tunnels dug into the mine's flanks, which followed the rich uranium ore veins. "The general level of radiation in the tunnels ranged from 1.5 to 2.0 mr [milliroentgens] per hour. Immediately adjacent to the face the radiation level was from 5.0 to 10.0 mr per hour for Gamma radiation and this increased to 15 to 20 mr per hour, when both Gamma and Beta were considered."

Doyle said fifty-three of the fifty-six people employed at the Monument No. 2 mine were Navajos. His readings showed that the Navajo workers were being exposed to from twice to nearly ten times the allowable amount of radiation by today's standards. In the worst cases, they were exceeding allowable weekly doses in less than one day and were reaching total annual doses in just a week.[3]

Doyle reported that a couple months after he made his initial visit to the Navajo Reservation mines, radon samples were found to be as much as 750 times the generally accepted limits, even by 1950 standards! "On January 11–12, 1950, an engineer from the Phoenix Office of the U.S. Bureau of Mines, Mr. Ernest R. Rodriguez, took radon samples in the West and East Reservation and in Monument Number 2. These radon samples were taken in accordance with the technique of the National Bureau of Standards and in containers furnished by that Agency. Those samples indicated a concentration of radon from 4 to 750 times the accepted maximum allowable concentrations of 1×10^{-8} curies per cubic meter."[4] (A curie is the amount of radioactivity in one gram of radium. It is 37 billion disintegrations per second for any atom.) Despite the clear and documented violation of safety standards, there is no mention of possible punishments or fines.

Doyle also took extensive measurements which showed that the air was thick with dust particles. In his report he confirms that miners were sent into the mines shortly after blasting, exposing them to excessive amounts of uranium dust, radon, and silica.

> Samples taken in the King Tutt No. 2 indicated that there is a concentration of approximately 20 million particles per cubic foot (mpcf) while mucking. [Loading the ore car with the broken rock. The accepted standard for dust at the time was 5 mpcf and was never to exceed 20 mpcf.] During wet drilling operations, a concentration of 6.4 mpcf was obtained in the vicinity of the driller. A dust sample was taken from King Tutt Number 2 tunnel 40 minutes after blasting when the tunnel had been reopened for work. The odor of [blasting] powder fumes at this time was still quite strong and the dust concentration was found to be 198 mpcf. None of these mines had been provided with forced ventilation and at the time of this visit no ventilation holes had been provided; no cross cuts had been made so that the natural ventilation in these mines is very limited.[5]

The Monument No. 2 mine in Cane Valley was also sampled for dust content, but because it was a naturally much wetter mine, dust did not present the same kinds of problems.

The general working conditions at the mines were abysmal by any standards, even excluding the excessive radiation and dust that Doyle observed. Clean, uncontaminated drinking water was not provided. In those mines that had damp walls, the miners drank water that dripped or leaked down the walls. No toilet facilities or places to change clothes or shower were provided. Such basic amenities would have greatly reduced the amount of radioactive dust that the miners carried home to their families. Doyle noted all of these problems, as well as the callous response he got when he suggested changes: "No change house, toilets, showers, or drinking water was [*sic*] available to the workers at these mines. It is my understanding that the U.S. Bureau of Mines made a safety inspection in June 1949, and recommended that a change house and basic sanitation facilities be made available at these sites. No action on these recommendations has been taken by the Vanadium Corporation of America and it is my understanding that they have doubts that the facilities would be used by the Indians if they were provided."[6]

Nevertheless, Doyle recommended that bath and toilet facilities be installed, that blasting be done only at the end of the day so that the dust could settle, that ore piles be wetted to control dust, and that "ventilation . . . be provided by mechanical means in order to reduce the radon concentrations to a safe level." He also added that, "Dust respirators should be furnished and the men should be required to wear them in unusually dusty operations."[7] Doyle concluded by asking for another, more detailed study to determine health conditions at the mines. Such a study was done to document the dangers, but it was not until the late 1960s, when more than 200 miners had already died, that the federal government mandated safety standards that approached those Doyle knew were needed in the late 1940s. By then, the biggest uranium boom the world had ever known was largely over. Many of the mines had closed or been played out, and the miners had been sent home, many to die slow, painful deaths.

The controversy over the health effects of working in the uranium mines, which took nearly twenty years to resolve, was not surprising considering the amount of misinformation and the predominant belief among the mining community and government officials that uranium mining was not dangerous. In July 1949, the commissioner of Indian affairs wrote to the USGS about the toxicity of uranium mining, the specific concern being what was termed "uranium poisoning." The response was completely misleading. The district mining supervisor of the USGS wrote to the Office of Indian Affairs that, "So far as silicosis and uranium poisoning are concerned, they [other experts] were in agreement with the statement I made in my former letter, that there is no danger whatsoever to the miners from these two complaints and that if there were any such dangers, that it would be more likely to appear in the mills processing vanadium and uranium ores, as the uranium concentrates there are very high grade." The letter goes on to say, "With regard to the dangers of radioactivity, it was the consensus of opinion that there was practically none under the conditions existing in any of the mines on the Indian Reservation, and there would be no danger until the content of uranium in the ores mined was higher than any that have been mined on the Indian Reservation."[8]

Such attitudes were prevalent and were clouded with the sense of national emergency and paranoia that arose when the Soviet Union detonated its first atomic weapon, starting the Cold War. No one wanted to be perceived as slowing down the scramble to produce uranium needed for America's burgeoning atomic arsenal and its growing demand for energy.

Matters were complicated by the fact that no miners had yet died from illness related to conditions in the mines. The assumption was that because the concentrations of uranium were relatively low, the resulting exposure to radiation was equally small and of practically no concern.

The government knew, however, that exposure to uranium and its deadly decay chain was dangerous. Grotesque cases of radiation poisoning had been documented in the early 1920s when factory workers in companies that produced luminescent dials began to lose their teeth, jaws, and finally their lives. Some died of leukemia, which doctors attributed to swallowing the radium paint. Luminescent paint was made by mixing zinc sulfide, a phosphorescent material, with radium and thorium. Workers, mostly females, used small paintbrushes to apply the radioactive and phosphorescent paint and would lick the brushes to keep them moist and pointed.

A 1929 report by the U.S. Bureau of Labor Statistics blasted the unsafe practices that were discovered in about thirty-one businesses that had been producing luminescent dials since 1913. The business had mushroomed partly because of the widespread use of luminescent markings on equipment in World War I, which made luminescent wristwatches popular. By 1929, out of 253 workers, 23 fatalities—nearly 1 of every 10 workers—had been recorded. Another 19 cases of injured but still living workers were known. In addition, "Aside from the cases recorded, nine were found that show possible connection with exposure to radium, especially four of these in which the persons were found to be *radioactive* when tested with electroscopes."[9] The study noted that these statistics, which were very alarming, may not have included all the victims because the industry had a high turnover rate. Because these symptoms were not always recognized at the time as being caused by radiation, many cases may have been misdiagnosed and consequently unreported.

Statistics showed that most of the victims became ill after just three to six years of exposure. However, some of the symptoms did not appear until after the workers had left the job for as long as nine years and were employed elsewhere. "Since, in the known cases, the disease did not appear for a considerable length of time, there is no way to ascertain how many of the workers, who had left their employment and perhaps moved to other localities, may have been affected. The symptoms were not specific, often resembling those of other diseases, and might easily be diagnosed according to the most prominent features. The disease is rare, and consultations with physicians proved that many of them were not familiar with it."[10]

The situation encountered by the radium dial industry was similar to the experience of the uranium mining industry some twenty years later. Like the dial painters, miners were a highly mobile group, working at mines throughout the Southwest for various lengths of time. This was especially true of Navajos, who according to mine operators, would work for several months and accumulate money, then disappear for weeks or months before returning. In addition, health problems suffered by miners that were later diagnosed as cancer were often misdiagnosed or obscured by the fact that the miner had smoked.

By 1929, the U.S. Navy, which used luminescent dials on ships, had already taken precautions in handling these materials and was well aware of the dangers. The navy regulations governing dial painting stated, "Radioactive materials are considered poisonous because if an individual is exposed to high concentrations for long periods of time, there will result disease processes which may become fatal." In the case of dial painting, the navy warned, "Exposure is principally through the mouth by swallowing and the inhalation of gases highly laden with the particles."[11]

The lessons learned in the 1920s had not been forgotten by 1940 when the United States was gearing up for war, and following the advice of an expert (Dr. Robley Evans of the Massachusetts Institute of Technology), the acceptable exposure standard was set at 10 picocuries per liter. (A picocurie is one-trillionth of a curie.) This was adopted by the National Council on Radiation Protection.

When the AEC officially became operative on January 1, 1947, it subsequently adopted this same standard for all its plants and laboratories. However, the AEC considered uranium mines and mills exempt from the close scrutiny and tight standards that had already been set for all the other facilities and plants which contracted with the AEC. From the moment people like Doyle arrived at the uranium mines in the West and found deplorable working conditions and high rates of exposure, they were forced to fight an uphill battle to get working standards for the mines set at just 100 picocuries per liter of air. This is ten times higher than the standards for other AEC operations. Despite the dangers to miners that were recorded and documented by Doyle and by Duncan Holaday, who assisted Doyle in monitoring the miners' health, the AEC never accepted responsibility for the health, safety, and welfare of the uranium miners.

Although the radium dial painting industry caused many deaths, some gruesome, nothing had yet been documented in the United States that demonstrated similar dangers for uranium miners. The previous problems had involved the use of radium, not uranium. By comparison, uranium is nearly a thousand times less radioactive than radium. In fact, much of uranium's radioactivity comes from the minute amounts of radium that occur naturally with uranium.

A source of the resistance to setting standards for uranium mining may have been a 1944 report by the National Cancer Institute on the most important direct comparison that American officials had: European uranium miners. Titled "Radioactivity and Lung Cancer: A Critical Review of Lung Cancer in the Miners of Schneeberg and Joachimsthal," and prepared by Egon Lorenz, the thirteen-page document was published by the U.S. Public Health Service, the National Institutes of Health, and the National Cancer Institute. It contained summaries of the many studies that had been conducted on the health problems of the pitchblende miners in Europe and their disease called *Bergkrankheit*, or mountain disease. Studies of the mysterious disease revealed that lung cancer afflicted the miners and had caused 75 percent of the deaths among them. Initially the cause of cancer was not thought to be uranium, but arsenic in the mine dust.

Despite the instincts of the miners themselves as well as the doctors who treated and studied them that radon gas and radon daughters were causing lung cancer, Lorenz was not willing to make a direct connection between lung cancer and radon gas, although European scientists had already made the connection. They even went into the mines and took radon measurements. Some of the mines were showing radon readings that were 200 and 300 times the level that was eventually adopted by the AEC for all operations except mines. One mine, which had been labeled the *Siebenschlenhen*, or death mine, by the miners, showed readings 900 times the AEC standard. However, Lorenz felt that a direct connection between radon and lung cancer could not be established because there were so many other complicating factors. For example, American health officials in the 1920s were stymied by the case of one luminous dial painter who had ingested so much radium that her breath contained radon equal to 10 picocuries per liter, or the maximum allowable safety standard that was later adopted. Scientists mistakenly concluded that radon could not be the cause of her death because the woman did not die of lung cancer but of bone cancer caused by radium poisoning.[12]

The problem faced by scientists and medical people in studying the health effects of radon and radioactivity is a problem endemic to epidemiology, or the study of the incidence and distribution of disease in a population. Without being able to isolate all the factors involved, it is often difficult to establish the cause of a disease.

Miners on the Colorado Plateau faced many of the same conditions as their European counterparts. The samples of radon, dust, and radiation that were taken in late 1949 and early 1950 by Doyle and Ernest Rodriguez from the Bureau of Mines in Arizona had been preceded by a flurry of activity. Ralph Batie, the chief of health and safety for the Colorado Raw Materials Office of the AEC, realized that any rapid expansion of uranium mining and milling in the West would bring with it dangers—dangers that were well known to health officials who had any experience at all with radioactive materials. He had been involved in sampling the air and exposures of mill workers in the Manhattan Project's clandestine program to extract uranium out of the vanadium mill waste piles. After the war the mills were closed, and he had been sent to other AEC plants. Now he was back to develop a health and safety plan for the coming uranium boom that was being prepared by the AEC.

Batie's first step was to get some idea of the existing conditions in the mines and mills. To add strength to his findings, he called in high-powered help from Washington and New York—Dr. Merril Eisenbud, head of the AEC's health and safety lab in New York, and Dr. Bernie Wolf, head of the AEC's medical division. From June 2 to June 5, 1948, they sampled the conditions of several mills in the area and one mine, "to evaluate the extent of the potential health hazards involved in existing and contemplated operations" so that "a suitable medical and engineering control program" could be established.[13]

This objective was not realized even though initial findings by Wolf and Eisenbud showed that there were serious health and environmental hazards in the mines and the mills. Wolf said that he and Eisenbud also intended to "define the type of studies which will be required in order to fully evaluate these hazards."[14] The two stated generally that, "The health problems which must be considered are primarily occupational, although some attention should be directed towards the matter of control over waste material and the spread of radioactive dust to areas adjacent to the mills."[15] Again, these recommendations were not only ignored but were grossly ignored. It wasn't until thirty years after this initial report that Congress approved a bill to clean up mill wastes.

The doctors found that radiation levels inside the mills were about 100 milliroentgens per *hour*, or about the amount that an average person is exposed to in nine months. The amount of radioactive dust in the mills, which was inhaled constantly by mill workers, was "considerably in excess of the preferred upper limit for safe working conditions." They found this dust especially high where the dried yellowcake was packed in barrels. They also reported that, "A potential health hazard from radon may exist in the mines and crusher houses of the mills."[16]

However, the mills and mines were not the only sources of contamination. Concentrated uranium was being blown all over the land surrounding the mills, the doctors discovered. "At Naturita a pronounced yellow staining of the rocks for a radius of a half a mile or so from the plant was demonstrated.... Rough calculations show that the amount of uranium thus deposited may be as much as 20 tons."[17]

The doctors recommended that all workers in mines and mills be given extensive and frequent medical examinations to monitor them for cancer, and that the small-town hospital at Uravan be supplied with the latest equipment and staffed by a full-time AEC doctor. They recommended that dust in all operations be limited to 10 million particles per cubic foot, that radiation in the air be severely limited, and that radon be limited to 100 picocuries per liter. Unfortunately, radon controls at this level were not imposed for another twenty years. Wolf and Eisenbud concluded by offering their support and expertise for any future consultation.[18]

After their report, Batie found that the AEC Raw Materials Division and other AEC officials did not want to hear about any extensive health program or anything else that might slow or, worse yet, derail the AEC's plans to develop U.S. uranium. They only wanted as much uranium as could be possibly found and as quickly as possible. After all, AEC officials argued, the states had traditionally been in charge of mines and mining. Let them worry about conditions in the mills and the mines.

Some thirty years later, Batie recalled the early days. "Both Mr. Leahy [manager of the Colorado Raw Materials Office] and myself received lots of static from the Washington Raw Materials Division headed by Mr. Jessie Johnson and his mining engineer, Mr. Gallagher—neither man was interested in the problems of Radiation Safety. In fact I was told to stay out of the mills and mines as far as recommendations for Safety was [*sic*] concerned. Also it was a State Problem."[19]

Batie then turned to the state of Colorado health officials, who were saddled with the problem because the AEC refused to accept responsibility for conditions in the mines and mills. Two state officials were cleared to test the Naturita mill owned by the Vanadium Corporation of America. Batie's supervisors allowed him to accompany the Colorado mining officials, but he could not take any samples. In 1949, the state's test results were released and were the same as those obtained by Wolf and Eisenbud.

Somehow the report on the mill and mine conditions, including the high levels of dust and radioactivity, was leaked to the *Denver Post*, which gave the story great play. Shortly after, Batie was asked to attend a meeting with AEC and Colorado mining officials and the VCA general manager, Denny Viles, who was incensed. Batie was told not to open his mouth at the meeting, where he and the state officials were harangued by Viles. Afterward, Batie was told his travel allowance was canceled. He could not leave the office. Early in 1950 he was granted his request for a transfer to AEC operations in Idaho. "I requested [a] transfer after trying in vain to operate a Health and Safety Program in Grand Junction and being told that the Director of Raw Materials in Washington wanted me fired for stirring up radiation problems in the mines and mills."[20] Batie's position as chief of health and safety for the AEC's uranium program for the entire Colorado Plateau remained empty for more than a year.

In August 1949, Dr. Roy L. Cleere, head of the Colorado Department of Health, appointed an advisory panel of federal, state, and uranium industry officials to oversee a comprehensive study of uranium mines and mills. The group agreed that little or nothing was known about the conditions and dangers in the mines and mills. A formal request for funds was sent to the Public Health Service's Industrial Hygiene Division on August 30, 1949.

While the Colorado officials waited for a reply, Henry Doyle, the ranking Public Health Service official in the region, had gone to work collecting samples from some of the mines on the Navajo Reservation where he knew he could obtain ready access. It was clear the AEC would do nothing in the area of health and safety, so Doyle knew that if anything were to be done, it was up to him and anyone else he could persuade to help him at his small outpost in Salt Lake City. Doyle was also aware that the mining and milling companies such as VCA and U.S. Vanadium, Climax, and others were increasingly resistant to health officials' claims that the mines and mills were dangerous. In an August 8, 1949, letter to Public Health Service officials in Washington, D.C., Doyle said that the adverse publicity that VCA had received over the conditions in its Naturita plant had put health officials on the outs with VCA. "As a result of this, I sincerely doubt that we shall be able to obtain much cooperation from VCA," he added. After meeting with a variety of Colorado State officials, Doyle said, "it appears that VCA has refused to make any of the corrections suggested in Mr. Jacoe's [the Colorado health worker who leaked the findings] report and if these alterations are to be made, it appears that the company will have to be forced into doing so by one of the state agencies."[21]

This reluctance of the AEC or federal health officials to take responsibility for the mine conditions resulted in a field day of unregulated behavior by the mining and milling companies. The effects of the AEC's refusal were compounded because most mining people refused to believe the mines were dangerous, and state regulators, who were suddenly told to do something about the mines, usually did not have the power to act. Doyle described the situation: "It has been determined that the State Bureau of Mines is the only agency having jurisdiction over this problem. The Director of the Bureau of Mines is very reticent about making any moves in this direction as he is not convinced in his own mind that uranium and radioactive ores represent a hazard to the health of workers."[22]

Mining officials and others felt that uranium mining was no more dangerous than other forms of mining and that this inherent danger was part of the job. Samples taken from mines all over the country revealed that radon existed in most underground mines to varying degrees.

Doyle quickly realized the extent of the problem he faced and set about getting help. He called on an old friend, Duncan Holaday, a fellow PHS employee and an expert in the field of radiation. Holaday could handle himself well with medical experts, politicians, and industrialists. He was not afraid of a fight and had the tenacity of a bulldog. After a phone call, Holaday helped Doyle immediately by providing sampling equipment and instructions, enabling Doyle to collect samples and readings from the VCA mines on the Navajo Reservation. Six months later, when the PHS approved a grant of $25,000, Holaday accepted the invitation from Doyle to come out to the Salt Lake City office, called the Occupational Health Field Station, to run the proposed comprehensive health study on uranium mines and mills. The grant went to the Colorado officials, who in turn agreed to work closely with Holaday. The full-scale study was launched in the spring of 1950 under Holaday's direction.

The study extended over two years and was broken into two phases—one on the health of the workers and the second on the working conditions in the mines and mills. Holaday, accompanied by a team of specialists, hit the road. The examination teams consisted of two physicians, two technicians, and an advance man who would contact the mine operators and make the necessary arrangements. As much as possible, the examination trailers, which doubled as sleeping quarters for the teams, were hauled to the mine sites. From July to October 1950, the team visited dozens of mines and gave free medical examinations to about 700 workers. In the second phase a year later, 460 more workers were examined, about 200 of whom were mill workers. In all, about 1,100 people were given physical examinations. The medical examination consisted of a physical checkup, a chest X-ray, and samples of blood and urine. The mines were tested for concentrations of radon and dust so that the readings could be correlated with the health results. Eight mills were studied, and forty-eight mines were sampled.

In May 1952, Holaday released his findings in a study titled "An Interim Report of a Health Study of the Uranium Mines and Mills." Not surprisingly it stated, "At this time, no clear-cut etiologic or pathologic patterns have been uncovered among the workers examined."[23] Holaday and Doyle knew that the majority of the workers had been in the mines less than three years and that in Europe it had taken twenty years for most miners to develop cancer. Nevertheless, they realized they had gathered valuable information that would serve as a basis for further health investigations.

Despite the disclaimer, the researchers were not finding perfect health. They found respiratory problems evocative of those of the European miners. The mouths of mill workers showed "a yellow coating on the tongue and teeth."[24] This was concentrated uranium oxide, which inspectors found floating freely in the mill air. It was so thick that it coated the mouths of mill workers. Clearly, yellowcake was also being swallowed and breathed, so it coated the miners' lungs as well.

Ben Jones is a Navajo who worked in the uranium mines for eleven years beginning in 1956 near Naturita, Colorado. He was responsible for drilling the holes into which dynamite was inserted and exploded to fracture the rock for removal. In a February 1993 interview at the Mexican Water Chapter House, he said, "The dust stayed in the air a long time. You could smell the gunpowder. When you blew your nose, it was yellow dust.... Sometimes they had vents, but it was not strong enough to move the air out." When the generator that drove the ventilation fans ran out of gas, "they just left it off."

George Brown, a Navajo, worked in the Tuba City uranium processing mill from 1959 to 1967. In a May 1993 interview, he said one of his jobs was to help pack fifty-five-gallon drums full of yellowcake. He had to stand beside the drum when it was on a vibrator that compacted the dust but also created choking clouds of yellow dust:

> We tried to put 1,000 pounds in a barrel. There were vibrators to pack it and that created a lot of dust. They would test our urine. But they never gave us the results. They would vacuum the air and test it, but we were never told [the results]....
>
> There were fumes in there. It really stunk. There was no ventilation. This was a danger, but no one ever told us at the time. After I made shift foreman... for about six years I worked with the really hot stuff.
>
> I used to come out with yellowcake all over my clothes and shoes. My kids played with the shoes and shoelaces. My daughter got cancer of the lymph nodes. My sister used to do all the laundry. She came down with cancer and after treatment had a relapse.

Doyle and Holaday found a higher than average incidence of pulmonary fibrosis—scarring of the lung tissue—which can develop into silicosis.[25] This, in its causes and symptoms, is similar to the black lung disease that has afflicted coal miners. The miners start coughing and develop phlegm, and in many cases, emphysema and severe shortness of breath, making it difficult to breathe and move around.

Some mill workers also had up to 60 micrograms of uranium in their urine samples. The yellowcake dust they inhaled and swallowed was making the workers radioactive from the inside out. Their internal doses of radiation were more dangerous than the external radiation they experienced daily. "It appears that the median value [of uranium in the urine] for mine workers is in the neighborhood of 2.5 micrograms and for mill workers 4.2 micrograms."[26] Holaday recognized that this was a major but unresolved problem, then as well as today. Inhalation of radioactive dust had long been recognized as a health hazard because uranium and radium, once inhaled, migrate to the bone marrow and remain there for their normal half-lives.[27]

The radon findings were equally alarming. More than 130 samples were taken in the forty-eight mines. The average reading was 3,100 picocuries per liter of air, which was 310 times the accepted nonmine standard of 10 picocuries and 31 times the suggested mine standard. The highest finding was 80,000 picocuries, which was 8,000 times the accepted nonmine standard. In the European mines the average was 1,500 picocuries, which was 150 times the nonmine standard and 15 times the suggested mine standard. The external doses of gamma radiation that the researchers found ranged from 25 to 176 milliroentgens per day. A normal range for the average American is from 125 to 300 per year.

Early in the investigation, Holaday and others realized that the internal doses of radiation that the miners and millers were receiving were the biggest concern. Experimental work by Dr. William F. Bale of the University of Rochester and the Health and Safety Branch of the AEC had shown that the decay products of radon could be concentrated in the lungs and produce much higher doses than had been calculated previously.

Holaday had been experimenting with the methods and costs of ventilating the mines and showed in his interim report how high levels of radon could be reduced dramatically simply by forcing fresh air into the mines. Because the radon concentrations were so high, Holaday and Doyle felt that it was necessary to temporarily put aside their full-scale environmental investigation and concentrate on controlling the radon in the uranium mines. In June 1951, Bale, Holaday, and Doyle worked on evaluating the problem and set up a series of conferences with mining companies in all the uranium mining states to discuss the problem. The meetings were held in mid-August 1951 in Salt Lake City. One conference was attended by health and regulatory officials from five states, as well as federal officials and members of the Atomic Energy Commission. A second group included twenty people from eight mining companies and three people from the U.S. Bureau of Mines.[28] The word was spread that there were potentially serious health problems in the mines and mills. Holaday reported that the companies showed a general willingness to work on the problem. Some of the companies responded immediately, and after return inspection trips, Holaday found significantly reduced levels of radon.

However, the reduction was nowhere near what Holaday knew was needed. His measurements showed that the levels of radon could easily be reduced to 1,000 picocuries per liter, which was a two-thirds reduction of the average level on the Colorado Plateau. With more effort on the part of mining companies, the 100-picocurie level could be reached. Holaday recognized that he was fighting an uphill battle because the ventilation needed would cost the mine operators money. This meant that many of the small operations which employed only a few miners would more than likely ignore the problem rather than reduce their already slim profit margins.* The remoteness of many of the mining operations and their sporadic schedule, Holaday knew, meant that many miners would be left to work in unsafe and ultimately fatal conditions.[30]

Holaday stated flatly in the interim report that radon and radon daughter readings were found to be "too high for safe operation over an extended period," and that "the median level of radon concentrations in the mines of the Colorado Plateau is above the median levels reported in the European mines. . . ."[31] He reiterated that a standard of 100 picocuries was considered to be the threshold of danger to life and argued that if mechanical ventilation were installed in the mines, mine operators should have no trouble meeting this standard.

Holaday and the others involved with this study did their best to keep everyone informed about the results of their work, including the Health and Safety Branch of the AEC and the National Bureau of Standards. When the interim report was distributed in May 1952, hundreds of copies were sent to federal government agencies, mining company officials on the Colorado Plateau, and state regulatory officials in the Four Corners area.

Although the report and its findings were widely known, especially within the industry, the AEC was nervous about its release and what newspapers and magazines might do with it. The United States was still dependent on foreign sources, which the AEC feared would be jeopardized. Most of all, the AEC was worried that if information on the dangerous conditions detailed in the report reached the miners, it might cause a general panic and a mass exodus from the mines.

In a summary report dated April 4, 1952, the AEC said, "There is no doubt that we are faced with a problem which, if not handled properly, could adverse-

---

*The amount, however, would be insignificant for the large companies—in a 1981 deposition, Holaday quoted Union Carbide engineers who put the cost at between 50¢ and $1 per ton of mined ore.[29]

ly affect our uranium supply. All information so far obtained indicates that practicable means exist to correct conditions which may be detrimental to health, but it is more difficult to avoid the distortion and exaggeration of fact(s) which are inherent in the subject."[32]

By 1955, mining companies were well aware of the radon problems in the mines and the fact that ventilation could be used to reduce those and other dangers. A Bureau of Mines circular on Kerr-McGee operations in the Lukachukai Mountains demonstrated with diagrams how mining companies could take advantage of natural as well as forced ventilation. The Bureau of Mines also spelled out the dangers: "Air impurities in uranium mines include radioactive dust and gases, gas and smoke from blasting, dust from mining operations, and exhaust gases when diesel-powered equipment is used. These impurities may be irritating, toxic or both."[33]

The circular noted that the Public Health Service was studying the health of uranium miners. It explained in detail the origin of radioactive material in uranium mines and how and why radon gas was dangerous. "Radon gas and dust contaminated by its daughter elements inhaled and carried into the lungs emit alpha particles in their process of decay with a potential hazard to lung tissue. Radon gas is dispersed into the mine atmosphere by diffusion from uranium-bearing rock and is released in drilling and blasting."[34] The circular also noted that the limit for radiation exposure was 100 micro-microcuries per liter of air (an amount that is equivalent to 100 picocuries) and stated that "proper ventilation and dust control will reduce hazards resulting from inhalation of radon and its daughter elements, reduce external radiation from radioactive compounds, and reduce hazards from possible chronic vanadium and uranium metal poisoning.... It should be mentioned that some mine water on the plateau contains high quantities of dissolved radon and its use in mining operations may result in an added source of radon in mine atmospheres."[35]

While the mine operators and government officials were well informed about the health hazards that workers in the mines and mills were exposed to daily, the miners were kept in the dark. Holaday and others were certain that the danger was extreme, but the obvious reluctance of state regulators and mine operators to believe this made the PHS study a delicate matter. The miners were not told of the dangers in the uranium mines until after the evidence had been accumulated.

Mining companies, of course, and the AEC did not want general panic created among the workers in the mines or the mills. Profits were at stake, the mining companies argued. National security was at stake, the AEC argued. That meant the miners and millers were left ignorant. In recalling the procedures of the study, Holaday said that the miners were never told anything unless there was an individual health problem noted after the examination. "Our procedure in the uranium study was to send the examinee a letter stating either that the findings were 'essentially negative' or one stating that the examinations showed possible problems...and that he should see his physician. His physician was sent a copy of the findings of the examination. The appropriate health officer was sent notices of cases of [tuberculosis]."[36] This meant that the miners were only informed of the "problem" in the mines after they had contracted a fatal disease.

Wilson P. Benally, age sixty-five, of Red Valley, worked in uranium mines for twenty-eight years, starting in 1948. He worked as a driller, a mucker (loading the ore cars), and at other jobs. He was part of the uranium miners' study and was examined periodically. "They did some tests two years back, and they found dust in his lungs," he said through his translator in February 1993. But "they never told him the results.

"So now the problem is that he [is] short of breath and sweats a lot, and he feels faint. He's under medication. He's using an inhaler now. They gave him some pill and told him to put it under his tongue. If it doesn't work, to go to the hospital."

Morris George, age sixty-six, worked in the uranium mines for six and a half years beginning in 1953 for Kerr-McGee in the Cove, Arizona, area. In a February 1993 interview through a translator, he discussed the health problems that he attributes to uranium mining: "He's breaking out in a rash that itches. He thinks it's from the uranium. He had an X-ray two years ago, but they never told him the results so he figures he's OK."

Holaday was frustrated but tenacious in his efforts and continued to fight for uranium mining safety standards for another twenty years. Holaday was always convinced that it was the AEC's and not the states' responsibility to ensure safe working conditions in the mines. "National security did not enter into the AEC's decisions at all, just a refusal to recognize a clear obligation," Holaday wrote later.[37] "The AEC was obviously trying to dump the problem on the state(s) while still trying to direct control [i.e., ventilation] work so that the industry would not be bothered. I think the best word to describe this activity is reprehensible."[38]

Holaday later became a key witness in lawsuits filed against mining companies and the federal government by the victims of radiation exposure in the mines. He felt strongly that "some measure of justice can be given to the families of those miners who have died as a consequence of their needless overexposure to radioactive substances in uranium mines. These unfortunates were as surely victims of the Cold War as those who died in Korea—but the official agencies, state and federal, have largely washed their hands of them and disclaimed responsibility. It is a sorry record!"[39]

Life, Not Death
Regulations Are Finally Established

THE ATOMIC ENERGY COMMISSION did not want to hear about health problems in the uranium industry and neither did the mining companies. State and federal health officials were left talking to themselves. Holaday's attempts to warn officialdom of the coming crisis were ignored and allowed to devolve into an endless round of studies and reports.

By 1953, nearly 800 mines on the Colorado Plateau were producing uranium on an intermittent basis. Some were open pit, but 157 mines employed at least 750 miners, most of whom were working underground. Holaday's latest information showed that 85 percent of all the underground miners were being exposed to more than 100 picocuries of radon, the recommended upper limit. Seventy percent of these miners were being exposed to levels between 500 and 1,000 picocuries; almost 25 percent of this group were being exposed to levels from 5,000 to 10,000 picocuries; and 15 percent of these miners were being exposed to levels of about 10,000 picocuries, or 1,000 times the generally accepted limit that was recommended in Holaday's interim report.[1]

The Uranium Study Advisory Committee, which was ostensibly backing Holaday's work, decided that the study should focus on just 300 or 400 miners concentrated in the areas around Uravan, Colorado, and Monticello, Utah. The usual examinations would be conducted—chest X-rays along with blood and urine tests.[2]

LEWIS L. STRAUSS OF THE ATOMIC ENERGY COMMISSION TESTIFIES BEFORE A CONGRESSIONAL COMMITTEE ON JUNE 2, 1954.
PHOTO: AP/WIDE WORLD PHOTOS

REP. W. STERLING COLE, TOP RIGHT WITH CIGARETTE, MEETS WITH REPORTERS ON NOVEMBER 13, 1954.
PHOTO: AP/WIDE WORLD PHOTOS

Later that same year, it became evident that the AEC was keenly aware of Holaday's work and was nervously watching the progressive examinations of the miners and the conditions in the uranium mines. To cover itself, the AEC worked hard to dispel the notion that nothing was being done to protect the miners. On July 13, 1953, Lewis L. Strauss, the chairman of the AEC, wrote a detailed five-page letter to W. Sterling Cole, then the chairman of the House and Senate Joint Committee on Atomic Energy (JCAE), the congressional panel assigned to oversee all aspects of the country's atomic energy program. Cole wanted to know what was being done about the proliferating reports of dangerous working conditions in the uranium mines and mills.

Strauss reiterated for the congressman all the data that Holaday had collected in his interim report and added the latest information that Holaday had delivered to the advisory committee in February. Strauss made it clear the AEC was aware of the uranium problem, but also that it was being worked on. "The general conclusion can be drawn at this time—namely that high concentrations of radon and its daughter products are common in uranium mines," Strauss wrote. "Nevertheless, a study of four mines showed that forced ventilation is capable of reducing by large factors the amounts of these elements present in the working areas of mines. The samples taken for radioactive dust and gas definitely show that methods to control the amounts of these elements present in the air should be provided in uranium mines."[3] Strauss even acknowledged that the recommended levels of radon should not exceed 100 picocuries per liter!

Sounding very sympathetic to the problem, he wrote that, "In short, while conditions in a majority of the mines have been found to be such that it would be hazardous for a person to work in them an eight-hour day, five days a week continuously over a period of 30 or more years, it is believed that the exposure accumulated to date by the individual miners in the uranium mines has not been sufficiently great to have produced injuries."[4]

Strauss suggested that yes, it was dangerous, but no real harm had been done. However, he offered no clues about the future. By suggesting that solutions to the uranium problem had been found, he implied that the solutions were being implemented. "It can be seen from the above that the states are aware of the situation, and with the full cooperation of the U.S. Public Health Service and the Atomic Energy Commission, are actively engaged in addressing and correcting it."[5] Nothing could have been further from the truth.

Meanwhile, back in the real world, Holaday and his handful of health officials continued to keep an eye on the mines and mining companies. At Laguna Pueblo, fifty miles west of Albuquerque, Anaconda was beginning to develop what would become the Jackpile mine. Holaday was aware that "a rather high concentration of radioactive dust in the Jack Pile mine" was found by Bureau of Mines engineer Ernest Rodriguez. Holaday personally delivered a warning to Anaconda mining officials in Butte, Montana. His warning was heeded when Anaconda analyzed the cost of ventilating the mines in what they quickly realized was a huge deposit. The mine shafts could go on for miles. Anaconda sidestepped the radon and ventilation problem by stripping off 300 feet of dirt and gravel and converting the mine to an open pit that covered 2,700 acres.[6]

Holaday was constantly warning company officials of the dangers of uranium mining and milling, hoping to prod them into action. In September 1953 he wrote to Kerr-McGee Oil Industries about their plans to build a processing mill. "As you will be constructing a mill, I should like to point out that it is much better to design and build in dust control methods at the time the mill is erected rather than to try to add them later. In all the mills which we have studied we have found that local exhaust ventilation must be used at all dust [creation] points in the crushing system as the ore has a high free silica content as well as containing radioactive material."[7]

Concerns about health conditions in the uranium mines persisted throughout the region. In September Holaday also met with Navajo tribal officials to discuss the health problems apparent in uranium mines on Navajo lands. Holaday reported to the tribe that even though progress in reducing radon and dust was being made, "all of them [the mines] are still much higher than the levels which were recommended in the Interim Report." The monitoring of these mines would basically be left up to the Navajos, Holaday said, because "it is impossible for either the state of Arizona or the U.S. Bureau of Mines to visit these isolated areas frequently."[8]

On September 17, 1953, Holaday met with Denny Viles of VCA to push for mechanical ventilation, which was nonexistent in VCA properties on the Navajo Reservation. These mines had higher concentrations of radioactive material than most of the other operations and would have more problems of control. None of VCA's mining engineers apparently had any training in mine ventilation, and therefore much of their control efforts were nonproductive. Viles tentatively agreed to an information meeting in late November so his engineers could learn about ventilation.[9] Such apparent ignorance on the part of professional mining engineers is questionable. How could one of the largest mining companies on the Colorado Plateau be ignorant about mine ventilation, since it was used so widely in all other kinds of mines across the United States and the world?

While private mining companies were slow to cooperate, fearing that word on dangerous mine conditions would be leaked to the public, so far there were no casualties, primarily because not enough time had elapsed since exposure. At a December 1953 meeting, the Uranium Study Advisory Committee told Holaday to develop a "census" during 1954 of the miners already examined and to perform as many follow-up exams in 1955 as possible to establish a baseline group. In addition, he and the study's doctors were directed to collect health and environmental data on the mill workers since the number of mills and the levels of exposure had increased.[10]

However, the committee waffled on informing the miners and millers about radon dangers, and delayed work on a bulletin on radon levels in the mines. At issue was the statement that more than 100 picocuries per hour might be dangerous. The figure was widely accepted among health officials and others as the acceptable threshold for radon in mines, but it had never been adopted formally by anybody. As a token concession to the problem, the AEC agreed to such things as a "special study" to see how fast radon gas oozed out of uranium ore. The committee agreed that this study was "a valuable contribution," then went on to more relevant reports, such as the fact that the American Standards Association had come up with suggested limits for worker radiation exposure but had avoided saying that these standards should apply to uranium mines.[11]

During the summer of 1954 more than 1,350 men from more than 250 mines were examined. Besides receiving the usual examination, each man provided detailed biographical information so that the miners could be followed over the years as they moved from mine to mine. Again, however, only 20 men had worked with uranium more than thirteen years, and 70 percent of the miners and millers had less than three years of experience.

One of the main ways that Holaday's teams followed the miners' lives was by poring over the obituary columns in the many small local newspapers in the region. While the AEC refused to accept responsibility for conditions in the mines and the health of miners, it nevertheless prepared for the inevitable results and arranged with officials in Colorado and Utah to autopsy every dead uranium miner. At least two groups of pathologists, one in Utah and one in Colorado, were to perform studies on all miners coming to autopsy and dying from all causes. Tissues from the lungs, liver, and other vital organs would be collected and studied for radioactivity levels. The AEC was finally going to establish a database on uranium miners.* The chief of the federal government's Occupational Health Division, Seward E. Miller, even appealed to the World Health Organization to conduct similar studies on miners in Africa, Europe, Australia, and Canada.[12]

In early 1954, in an attempt to ward off criticism that it was doing nothing to improve conditions in the mines, the AEC finally released a public document on the environmental hazards of uranium mining. Titled *Control of Radon and Its Daughters in Mines by Ventilation*, the pamphlet was the result of work that had been done earlier by Holaday and others on how much fresh air was needed to blow out the radon and reduce radon levels in the mines to acceptable standards. For a dead-end tunnel, it recommended 500 cubic feet of air per minute, and for larger rooms, up to 1,000 cubic feet of air per minute. The pamphlet warned, however, that "Maximum use should be made of natural ventilation, but this cannot be relied upon to bring concentrations in dead-ends down to acceptable levels even under the best of conditions." However, since there was no oversight or enforcement involved, the information was noted and generally disregarded.[13]

* * *

In November 1956, a fifty-one-year-old uranium miner named Tom Van Arsdale died of lung cancer in Grand Junction, Colorado, leaving a wife and several small children without a source of income. The autopsy showed oat cell

*The study was found to be impractical and was never fully implemented. However, in 1951, an autopsy program was initiated to examine the lungs of deceased miners (personal communication from Dr. Victor Archer, February 9, 1994).

DR. VICTOR ARCHER WAS IN CHARGE OF THE URANIUM MINERS' HEALTH STUDY DURING MOST OF THE 1950S AND 1960S.

carcinoma of the left lung. This was the same form of cancer that had decimated European pitchblende miners.

Van Arsdale had been part of the PHS uranium study, having been examined on August 4, 1953. He had been in and out of the uranium mines since 1940. Analysis of his urine showed 6.8 micrograms per liter. He had had regular chest X-rays beginning in 1940. Silicosis of the lungs appeared in his 1953 chest X-rays, and by 1955 the condition was severe. Lung cancer showed up the next year—sixteen years after his first exposure to uranium.[14]

A year later, in the fall of 1957, Van Arsdale's widow filed in Colorado for workmen's compensation. She was represented by an attorney who worked for the same law firm that represented the former U.S. Vanadium Corporation, which had changed its name to Union Carbide Nuclear. Victor Archer, the young doctor who was in charge of the field operations of the PHS study, was asked to testify at the hearing. In a lengthy memo about the case, Archer explained that the widow's attorney knew that an award from the state of Colorado was unlikely because a similar claim in 1954 had been turned down. The attorney wanted only a "maybe" ruling from the Workmen's Compensation Board so that the case would be left open until Archer's study turned up something positive. Archer was being asked to testify about the continuing PHS study and the medical importance of the uranium and other radioactive elements found in Van Arsdale's body. "It is not likely that a better documented case will come along for some time," Archer wrote. He knew that by itself the case did not prove much scientifically, but that it was a very strong first case. "In other words, a case which suggests a relationship but does not prove one."[15]

Archer could not testify, however, without permission from his superiors in Washington, D.C., who had to balance the gains to be made from getting documentation on uranium and miners' deaths on record against the likely reaction of the mine owners, who might become alarmed at the adverse publicity and refuse to cooperate with personnel conducting the study on the miners. In the end, Archer was allowed to testify, arguing that if he did not, the PHS would be disregarding its responsibility for the health of miners. His testimony caused no reaction, and the studies continued unimpeded.

By the end of 1957, a total of 3,200 miners had been examined since the beginning of the study, but only about 500 had been examined more than once because of the high mobility of the miners. The examiners made an effort to obtain as much personal background information as possible so that work histories and follow-ups could be related. During the intervening summers, the members of the PHS office had sent out questionnaires to all those who had been examined so that contact would not be lost. Archer said that they had kept in touch with 97 percent of all who were examined.

Miners from 290 separate mines were examined that year. The medical examination had been expanded to include rectal exams, lung capacities, and studies of hair roots as well as "sputum" smears. These sputum tests were based on a technique similar to the "Pap" smears of today. Samples of the mucus from the miners' lungs were taken, then processed and examined. Doctors had found this was a quick and easy way to detect the presence of cancer cells in the lungs, and the results could serve as a warning for the advent of cancer. Of the 1,343 miners who received a sputum test, 14 were found to have possible indications of lung cancer, Archer reported.[16]

There were other worrisome signs. A greater number of the men were showing up with pneumoconiosis, the result of scarred lung tissue from inhalation of dust. In the study group, Archer reported that 61 deaths had been recorded, most from causes unrelated to mining. However, many of the miners had also died of diseases suspiciously related to their occupation. Eight had died of malignant tumors, and 17 had died of lung and kidney diseases. Three miners had died from infections, 1 had died of alcoholism, and 4 had died from suicide. Twenty-eight had died in accidents, about half of which occurred in the mines, and others from auto accidents. Four deaths were from unknown causes. Still, Archer was cautious. He would not yet say that uranium was killing the miners.[17]

Meanwhile, word about the condition of the mines and the miners was trickling into Washington, D.C. On March 11, 1959, the Joint Committee on Atomic Energy conducted a hearing on the status of the uranium industry and asked Duncan Holaday to testify. His testimony was the first official recognition that the U.S. Congress took of the potential problem. Earlier, Holaday's former boss, Henry Doyle, who was in Washington, D.C., had testified about silicosis problems in the mines.

On the advice of his superiors, Holaday's testimony was cautious and almost noncommittal. His PHS bosses apparently were hoping to use this first hearing to educate rather than to raise an alarm. He testified that about 4,500 men were currently working in uranium mines on the Colorado Plateau, with the great majority working in the Four Corners states of New Mexico, Colorado, Utah, and Arizona. Another 5,000 people, he estimated, were working in crushing plants or refining mills. All, he said, "are exposed to external gamma and beta radiation and to internal radiation from inhaled radioactive elements which are present in working areas."[18]

Holaday painted a clear picture of the dangers that the uranium miners and millers faced daily, the history of the European pitchblende miners, and the mining companies' apparent lack of concern about the problem. He complained about the lack of legal muscle and willpower among state and federal agencies to impose and enforce strict controls on radon levels and radiation hazards.

However, late in 1959 some hope for possible strict controls surfaced. Having become aware of the legal and bureaucratic fog surrounding the issue of responsibility for uranium mine safety through concerns pressed by Holaday and Doyle, the Public Health Service called a federal interagency committee meeting. Attending were officials from the departments of interior; labor; health, education and welfare; and the Atomic Energy Commission.

The committee decided that the only clear responsibility lay with the Department of the Interior because of its responsibility for Indian lands and consequently the conditions of the mines on them. Otherwise, the committee said, the responsibility of the federal government was unclear. "Mines are excluded from the jurisdiction of the Atomic Energy Commission both by its licensing regulations and the Atomic Energy Act as amended."[19]

The AEC was falling back on its previous stance that even though it was the only legal buyer of uranium, it had no responsibility over the uranium until it was "removed from its place of deposit in nature." The phrase was very broadly interpreted by government officials as meaning that the government accepted responsibility for the uranium only after delivery of the ore. This seems more than a little disingenuous in light of the AEC's 1947 incentive program that provided a large bonus for ore received from a newly found mine. This bonus plan by itself would appear to indicate deep involvement in the mining end of the process.

One possible approach mentioned was that the U.S. Department of Labor had the ability to control and dictate conditions in the mines through the Walsh-Healey Act, a law passed in the 1930s that initially specified conditions for child labor. Over the years the law had been expanded and amended to cover government contracts for the purchase of goods and services. It allowed the government to ensure that the producer or manufacturer of goods purchased by the government maintained safe working conditions for its employees. This readily applicable law was considered by the committee but ultimately rejected "pending a legal opinion."[20]

The committee suggested that even if the legal authority to regulate mine conditions did rest with the labor department, the problem could not be solved because "the Department of Labor does not now have the technical staff needed to enforce the Public Contracts Act in uranium mines.... Enforcement of the Public Contracts Act would not give the mines the needed technical assistance to control the potential problem."[21] Ironically, in another seven years this same law would be used by the Department of Labor to clamp down on uranium mining conditions. The difference was a federal official who was willing to accept responsibility for ensuring the public welfare with which he had been entrusted.

The Department of the Interior's Bureau of Mines was considered as a possible controlling agency, but bureau officials pleaded that it had "no enforcement powers in mines located on private property." The Bureau of Mines could, at the recommendation of the U.S. Geological Survey, cancel mine leases to private operators on public lands if the operators did not comply with specific recommendations. In what was perhaps one of the greatest examples of bureaucratic buck-passing in American history, the collection of federal agencies unanimously decided the uranium mines were not its problem but in fact were the problem of the states in which the mines were located. "The Committee recommends that a conference of governors of the states concerned with uranium mining (Arizona, Colorado, Idaho, Montana, Nevada, New Mexico, Oregon, South Dakota, Utah, and Wyoming) be called to discuss the problem."[22]

In March 1960, Holaday, Archer, and others in the Salt Lake City office of the Public Health Service were gearing up for that summer's medical survey of the uranium mines and miners. The teams estimated that about 2,500 to 3,000 men were employed in the mines, along with another 5,000 men in the underground mines in the Grants, New Mexico, area and the expanding Jackpile mine at Laguna Pueblo. However, except for the mines on the Navajo Reservation in the Shiprock and Tuba City areas, miners in the Grants area would not be examined. The survey would be conducted by two medical teams of eleven members each who would be in the field for six weeks at a time. In addition to the usual X-rays and other examinations, the doctors would ask the miners about general lung conditions and such problems as shortness of breath and chronic bronchitis, which are indicators of silicosis. The miners in the potash mines near Carlsbad in southern New Mexico would be used as a control group for comparison.

While the medical teams were out in the field, the secretary of the interior issued a report on how radiation hazards in the uranium mines on Indian lands were being managed. It stated that, "To date, gamma radiation rates in excess of the prescribed limits have not been detected in uranium mines." The report, dated August 9, 1960, added that the department would not take a stance on whether the radon levels were dangerous until the Federal Radiation Council, which had been formed to develop such policies, would issue standards. The council would continue to meet on this issue without taking a position for another half-dozen years.[23] Meanwhile, mining conditions, as Holaday and other members of his group had shown time and again, would continue to be deadly for the miners.

On December 16, 1960, the governors of New Mexico, Wyoming, Utah, and Colorado assembled to discuss the uranium mine problem. Also present were representatives from Arizona, Montana, Oregon, and South Dakota, along with the Atomic Energy Commission and members of the departments of the interior; labor; and health, education and welfare. Everything known about the problem to date, including death rates, conditions in the mines, and the legal situation, was put together in a summary report prepared by the Public Health Service. The picture was as muddled as ever. Statistically there was still no evidence that the radiation in the mines was killing people. Clearly, however, mine safety did not present an attractive picture. Of the 3,317 miners who had been examined by 1960, 108 had died. Mine accidents were the biggest killer, causing one-third of the deaths. Heart disease was almost as bad, killing 27 percent. More than a third of these deaths were the result of complications from silicosis. Auto accidents and miscellaneous causes accounted for another 18 percent, and cancers, including lung cancer, accounted for another 10 percent. Suicide and other causes accounted for the remainder.[24]

The most telling statistic was a comparison of the health and death rates for uranium miners with those for the general population. The comparisons were frightening. Among the miners, death from heart disease was 11 times what it should have been. For miners with more than three years of experience in the mines, it was 17.8 times what was expected. Cancer of the respiratory system was 5 times the expected rate. The sputum test for cells suspected as cancerous indicated that the cancer rate increased dramatically from 1957 to 1960. In 1957, 1 percent of the miners showed positive signs of cancer cells. Just three years later, that figure tripled to 3.3 percent. If that rate continued, officials worried, nearly 30 percent of the miners would have lung cancer within the next decade.[25]

Conditions in the mines had improved, but were still far from safe, the governors learned. The entire system of determining a level of exposure had been simplified. Beginning in 1954, Holaday and his staff had set up 300 picocuries per liter of air for radon daughters as a "working level" of radiation in the mines. This level was later reduced to 100 picocuries per liter based on work done by Dr. William F. Bale in 1957. It meant that 1 "working level" was the amount of radiation to which a miner might be continuously exposed for an average of forty hours per week over thirty years without suffering any damage to his lungs.

GOV. STEPHEN L. R. McNICHOLS OF COLORADO WAS AN OUTSPOKEN CRITIC OF THE FEDERAL GOVERNMENT'S LACK OF ENFORCEMENT OF SAFETY STANDARDS IN URANIUM MINES.
PHOTO: AP/WIDE WORLD PHOTOS

The designation of a "working level" made it much easier to communicate the relative dangers of mine conditions to miners and other nontechnical people. If a miner was told he had been exposed to 2 working levels, it meant he had been exposed to twice the allowable level for any given period of time. The governors' conference readily accepted the terminology. Samples taken in 1958 showed that the mines in Arizona and many others still had high levels of exposure. Nearly half of the Arizona mines had more than 10 working levels, followed by Colorado, where 65 percent had more than 3 working levels. New Mexico mines, which employed about 700 miners, or less than a third of the total, reported an average of 1 working level.[26]

Samples taken in 1959 were offered for comparison with the 1958 readings but were confusing. Colorado mines had gotten worse, going from an average of 4.6 working levels to 6.1 working levels. Mines in Arizona had improved significantly, going from an average of 8.3 to just 2.9 working levels. In New Mexico, conditions had remained relatively stable.[27]

When it came to the issue of safety, the governors agreed that there was no dispute that a problem existed. However, the issue of who should or could do anything about the problem remained unresolved. The governors, led by Gov. Stephen McNichols of Colorado, railed at the federal government, claiming that states lacked the power to regulate activities such as uranium mining, which was controlled by the federal government. McNichols argued that the states lacked the financial ability to hire the people needed to regulate the mines. The AEC had started and was running the uranium mining program, not the states, and therefore should give the states the money to inspect and control the mines if the AEC was unwilling to do it. McNichols complained that no matter what the states did, the AEC would have ultimate veto power. More maddening to company officials who had mines in different states was the fact that each state had a different set of rules and regulations.[28] The conference ended with no clear resolution of the problem.

Three months later, the AEC, stung by the criticism it received from the governors, responded with a press release saying that in 1959 the U.S. Congress had passed a law that allowed the AEC to establish the conditions under which it could pass along its regulatory control over such things as uranium mines to the states. It let the states set up all the rules, but the AEC had veto power. The statement from the AEC was nothing more than another dose of the bitter medicine the states were unwilling to swallow.

A clear line of authority and responsibility had to be established if the endless bureaucratic buck-passing was to end. The clearest place to begin was the Bureau of Mines in the Department of the Interior, but first its responsibilities had to be expanded. On September 26, 1961, Congress ordered the interior department to study mine safety in all noncoal mines in the United States, including accident rates, health, safety, and training, and to survey all applicable state laws.

Five years later, on March 21 and 22, 1966, the labor subcommittee of the Senate Committee on Labor and Public Welfare held hearings on a package of bills that had been drafted to address a wide range of health and safety problems in the noncoal mines of the United States. Included in the bill was the problem of radiation in the uranium mines. Heading up the testimony on the health issues was Henry Doyle, still with the Public Health Service. After telling the committee about the serious silicosis problems in mines across the United States, he launched into details of the uranium miners' study still being conducted on the Colorado Plateau. By now, the evidence was unmistakable. Out of about 3,200 miners, including about 750 Indians, 60 had died from lung cancer. Doyle said:

> We have now 60 lung cancers appearing among our study group of some 3,200 uranium miners. We have now been able to construct a…curve that gives a time-dose [exposure-response] relationship. It demonstrates very dramatically that as the time exposures in the uranium mines go up then the risk of lung cancer becomes increasingly great….
>
> The miners who had the lowest exposure rates or those that approximate the working level which has been recommended had an incidence rate of 3.1 per 10,000 men. Those men that were in the highest exposure group had an incidence rate of approximately 120 per 10,000 men.
>
> So we find that the incidence rate varies from one of normal in the lower group to some 40 times normal in the higher exposure group. Of course, the intermediates fall within these ranges. The normal lung cancer rate for people living in the uranium mining area but not associated with uranium mines is 3 per 10,000 men.
>
> Since the governor's conference of 1960, there has been a continuous decrease in the number of men exposed to the higher concentrations. This is based on information which we collect during the last quarter of each year. The exception to this was in 1965 when the number of men exposed to higher concentrations began to go up. Now, whether this is an anomaly or whether this simply indicates a new trend, we are not prepared to say.
>
> It is true that as the mines become deeper and the smaller mines re-enter the uranium mining picture, then the control problems become more severe.[29]

Unfortunately, little notice was taken of these alarming statistics. The uranium miners' plight was buried in the middle of the other health and safety problems of noncoal miners across the country. Doyle's testimony on worsening conditions in 1965 showed that the radiation and radon problems in the uranium mines were not under control, despite what industry officials and others wanted people to believe. The bill that emerged from the hearings became the Federal Metal and Nonmetallic Mine Safety Act and was signed into law in late 1966. It placed mine safety under the control of the Department of the Interior, headed then by former Congressman Stewart Udall. The law sent hardly a ripple through the uranium industry. Specific regulations and standards would not be in place for several more years and only after lengthy hearings, comments, and appeals from affected parties. Then, the industry would be given a period of time in which to comply. Meanwhile, uranium miners were dying.

* * *

In early 1967, an enterprising reporter named J. V. Reistrup of the *Washington Post* heard about the lung cancer problem among uranium miners on the Colorado Plateau. He made a trip to Nucla, Colorado, and interviewed a fifty-six-year-old uranium miner named John Morrill, who was dying of lung cancer. Morrill began uranium mining in the early 1940s and continued until 1965. About ten years of his nearly twenty-five years in the mines were spent in the small mine operations called "dog holes" in which a handful of men would work with picks, shovels, and wheelbarrows. The dog holes, like most of the mines on the Navajo Reservation, were sometimes leased from the bigger mining companies such as VCA, Union Carbide Nuclear, and Kerr-McGee. Morrill fit the classic case of lung cancer found among the European miners. He had more than twenty years in the mines and was in his mid-fifties. He had lung cancer and knew he would be dead within six months. Despite the impending tragedy, he was fatalistic, accepting death as inevitable. "If I hadn't been working in the mines, I might have been here a lot longer—maybe; I don't know."[30]

Morrill had filed for $12,000 in Colorado workmen's compensation benefits due to him because of his disabling condition, but his claim was being fought by his former employer, Union Carbide Nuclear. Reistrup talked with Holaday and Archer, who gave him an earful. "The only thing that has saved us from a real calamity is that many of these people work only two, three, four years," Holaday explained. "There'll be more. We haven't seen them all."[31]

The article hit Washington, D.C., like a bombshell. Until then, no one had told the general public about the extent of the uranium mining problem in the West. Public fascination had been with the Cold War, bomb shelters, and nuclear energy and its possibilities. Now there was proof that miners were dying of lung cancer caused by radiation in the uranium mines. Already the dangers of

MANY SMALL MINES SUCH AS THIS ONE ON THE ROCKY, FLAT MESAS OF RED VALLEY, ARIZONA, WERE KNOWN AS "DOG HOLES" AND LACKED ADEQUATE VENTILATION.
PHOTO: MURRAE HAYNES

SECRETARY OF LABOR WILLARD WIRTZ AT A PRESS CONFERENCE ON NOVEMBER 21, 1962.
PHOTO: AP/WIDE WORLD PHOTOS

radioactive fallout from atmospheric testing had caused President Eisenhower to halt the tests because the fallout was killing American citizens.

The article also got the attention of Congress. On April 21, 1967, the Joint Committee on Atomic Energy issued a press release announcing a series of hearings from May 9 to August 10 on the issue of radiation exposure among

THIS URANIUM MINE ON TUTT MESA IN RED VALLEY WAS LEFT OPEN FOR ABOUT FORTY YEARS.

THIS “DOG HOLE” AND OTHERS LIKE IT WERE UNVENTILATED, ACCUMULATING LARGE AMOUNTS OF DEADLY RADON GAS.

URANIUM MINES WERE OFTEN DUG INTO SANDSTONE, WHICH IS UNSTABLE.

THE ENTRANCES TO SOME MINES WERE SUPPORTED BY TIMBERS.

This page and following spread

RADIOACTIVE MINE WASTE AND DEBRIS EXPOSED NAVAJOS AND FARM ANIMALS TO ADDITIONAL RADIATION.

A SCINTILLOMETER ON A PILE OF MINE WASTE SHOWS HIGH LEVELS OF RADIOACTIVITY.

URANIUM MILL TAILINGS LEFT EXPOSED AT THE HOMESTAKE MILL NEAR GRANTS, NEW MEXICO.

ONE OF MANY MULTILEVEL MESA MINES NEAR RED VALLEY THAT EMPLOYED HUNDREDS OF NAVAJOS.

CLOSEUP OF MESA MINE AND WASTE.

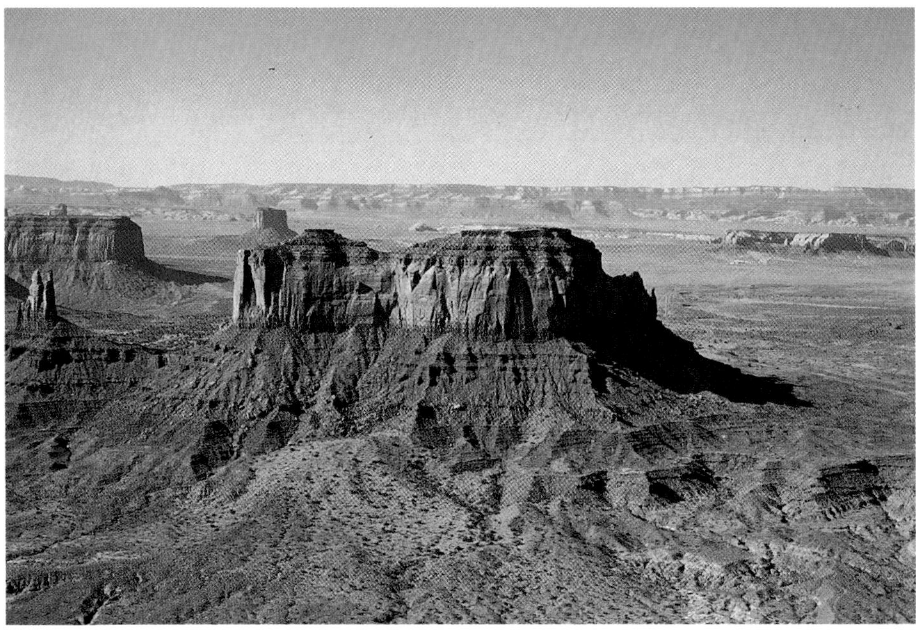

BUTTE IN MONUMENT VALLEY.

OLJATO MESA IN MONUMENT VALLEY
WITH URANIUM MINES AND WASTE.

RECLAMATION AT THE JACKPILE MINE INVOLVES EXTENSIVE GRADING OF SOIL.

WATER ACCUMULATION AT THE BOTTOM OF THE JACKPILE MINE IS CONTAMINATED WITH RADIOACTIVITY.

THE LIGHT-COLORED MATERIAL IN THE FOREGROUND IS LOW-LEVEL ORE THAT WILL BE COVERED WITH SOIL.

RECLAMATION MANAGER JIM OLSEN INDICATES A RECLAIMED MINE SITE.

MOONLIGHT MINE PRIOR TO RECLAMATION.

WASTE PILE IN FRONT OF MOONLIGHT MINE CONTAINS HIGHLY RADIOACTIVE WASTE.

RECLAMATION INVOLVED PUSHING MINE WASTE BACK INTO THE PIT.

BULLDOZING MINE WASTE INTO THE PIT.

AERIAL VIEWS OF THE MONUMENT NO. 2 MINE IN CANE VALLEY, ARIZONA, NEAR MONUMENT VALLEY.

MILL TAILINGS PILE IN CANE VALLEY WITH HOUSES NEARBY. MONUMENT VALLEY IS IN THE BACKGROUND.

LUKE YAZZIE, WHOSE HOME IS IN CANE VALLEY, IN THE 1940s BROUGHT GEOLOGISTS TO URANIUM DEPOSITS THAT BECAME THE MONUMENT NO. 2 MINE.

THESE TRUCKS TRANSPORTED THE MILL TAILINGS PILE FROM CANE VALLEY, ARIZONA, TO MEXICAN HAT, UTAH.

THE DEPARTMENT OF ENERGY TOOK CARE TO HOSE RADIOACTIVE SAND OFF THE TRANSPORT TRUCKS.

AERIAL VIEW OF MEXICAN HAT TAILINGS PILE NEAR THE SAN JUAN RIVER IN UTAH.

ROGER HATHALE, NAVAJO MEDICINE MAN, WITH HIS WIFE. HATHALE BLESSED THE TAILINGS REMOVAL PROJECT IN CANE VALLEY.

SHIPROCK MILL TAILINGS PILE NEXT TO THE SAN JUAN RIVER. RADIOACTIVE MATERIALS HAVE BEEN DETECTED IN THE FLATS BETWEEN THE PILE AND THE RIVER.

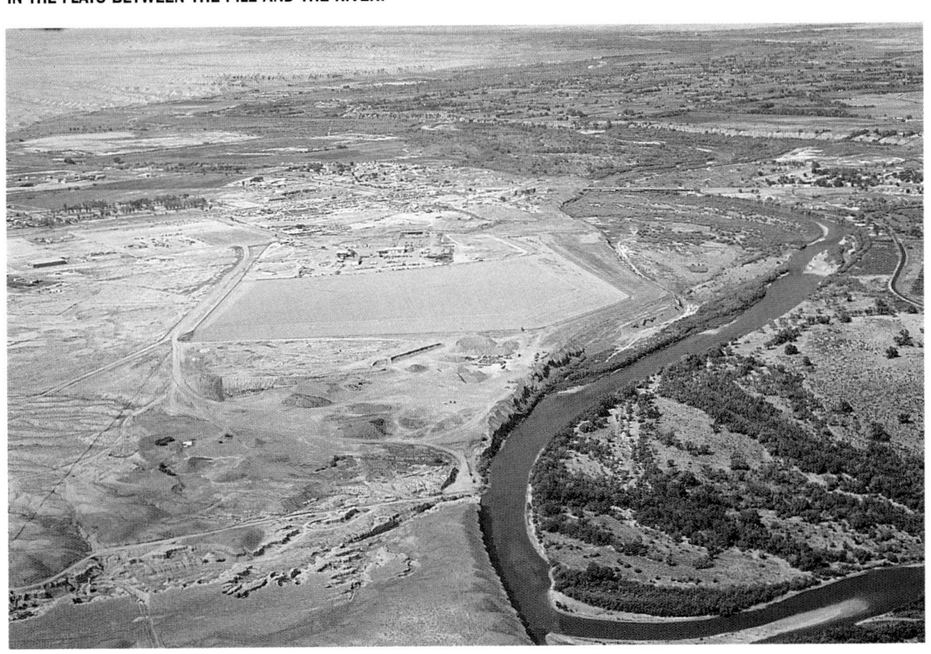

RETIRED DEPARTMENT OF ENERGY GEOLOGIST WILLIAM CHENOWETH VIEWS THE CONCRETE SEAL COVERING A FORMER URANIUM MINE IN RED VALLEY.

MILL TAILINGS PILES ARE GRADED AND THEN COVERED WITH CLAY AND ROCK BY THE DEPARTMENT OF ENERGY.

RECLAMATION MANAGER PERRY CHARLEY, LEFT, WITH WILLIAM CHENOWETH IN FRONT OF A URANIUM MINE WASTE PILE.

the uranium miners. Suddenly congressmen who had lorded over the AEC by virtue of their position on the committee were extremely concerned about the miners' welfare. "My primary concern is to see that all necessary action is taken to protect the workers," said Rep. Melvin Price, whose subcommittee had been conducting hearings on the issue for a half-dozen years. "My immediate concern is that the needed action—setting the limits and seeing that they are followed—is taking too long. In these hearings we plan to find out why it is taking so long to do this."[32]

The *Washington Post* article was brought to the attention of Willard Wirtz, secretary of the Department of Labor. A tough-minded administrator and life-long advocate of labor safety, he was horrified to learn what was happening. He talked the situation over with his associate, Esther Peterson, then the assistant secretary for labor standards. After a quick review of the situation, Wirtz learned that the department all along had had the power to impose safety standards on uranium mines under the Walsh-Healey Public Contracts Act. This was the same legal authority that the labor department rejected in 1959 during an interagency meeting that preceded the 1960 governors' conference. Irate at what he discovered to be an incredible history of bureaucratic buck-passing, he formulated a plan to impose a highly restrictive standard of 0.3, or one-third of a working level, for radon in the uranium mines. This standard equated to 100 picocuries, the level that Holaday and others had said would put uranium miners at no greater statistical risk than the rest of the working population. The new standard would apply to all mines providing ore to the Atomic Energy Commission. Wirtz was prepared to move swiftly and decisively, but he held off pending a May meeting of the Federal Radiation Council.

President Eisenhower created the council by executive order on August 14, 1959. It was given no authority but was to develop policy guidelines that individual agencies would adopt and execute. It included agencies that might deal with radiation: the secretaries of the departments of defense; health, education and welfare; commerce; labor; agriculture; and the chairman of the AEC. The council had produced an endless stream of studies but had been unable to reach consensus. The *Washington Post* article appeared just days before the council was to meet to establish radiation standards for mines.

When it met on May 4, things were no different. In a split vote, the council failed to adopt strict standards. Wirtz was incensed. On May 5, he issued his order and said all public contractors would be forced to meet the 0.3 working level standard. He opened up the standards to a thirty-day comment period, after which adjustments might be made and a final order issued.

Industry officials and politicians who considered the AEC their personal domain were incensed. Suddenly the members of the JCAE who just weeks earlier had said their biggest concern was the welfare of the uranium miners were now under attack for having done nothing as miners died from the well-

REP. CHET HOLIFIELD OF CALIFORNIA, ACTIVE IN ATOMIC RADIATION ISSUES, TESTIFIES AT A JUNE 19, 1959, CONGRESSIONAL HEARING ON ATOMIC FALLOUT.
PHOTO: AP/WIDE WORLD PHOTOS

known but uncontrolled dangers in the uranium industry. On the first day of the hearings, May 9, the *Washington Post* blasted the politicians and asked if they would be willing to sit in one of the uranium mines while they discussed what level of protection should be imposed. Representative Price began the hearings defensively, pointing to a list of hearings on the topic dating from 1959 and a stack of studies.

The first witness was Dr. Paul Tompkins, director of the Federal Radiation Council, who reiterated for the committee everything that had been done and discussed in the past years. What followed was a litany of reasons why no one had done anything.

Rep. Chet Holifield, a Democrat from California and an influential member, typified the prevailing attitude in Congress that because the issue was complicated and there was no certified proof that radiation in the mines killed miners, no action should be taken. When Tompkins said the council could not agree the day before on what to do, Holifield pounced on that statement:

> Outside of accusing one side or the other of selfish interest one way or the other, and certainly I do not accuse either side of that, you must come to the conclusion that they were dealing with a complicated problem upon which common agreement had not been reached.
>
> I have had too many scientists before me taking different positions to not understand that this can be done with honesty and integrity. It is a matter of evaluative judgment, of subjective judgment, and not judgment based on a common agreement as to facts.
>
> Is that not right?[33]

Tompkins, of course, agreed.

Congress and the barons of the AEC were embarrassed and upstaged by Wirtz, who had caught them in a morass of their own making, having catered too much to industry and big business while paying lip service to the little guys. Wirtz was scheduled to testify in the afternoon session on May 9, and the committee was eager for revenge. However, Wirtz was a tough-minded lawyer and politically savvy. He had Congress on the run because he had acted when

Radiation Exposure of Uranium Miners

Mortality summary by states and year*

STATES WHERE MINERS WORKED			
Colorado	**70**	Colorado/Utah	**7**
Utah	**12**	Colorado/Arizona	**1**
Arizona	**1**	Wyoming/Utah	**1**
Wyoming	**1**	New Mexico/Arizona	**1**
Colorado/New Mexico	**1**	New Mexico/Utah	**1**
Total	**96**		

TOTAL DEATHS BY YEAR			
1945	**1**	1956	**2**
1946	**0**	1957	**3**
1947	**1**	1958	**5**
1948	**0**	1959	**5**
1949	**1**	1960	**9**
1950	**1**	1961	**6**
1951	**1**	1962	**7**
1952	**0**	1963	**10**
1953	**1**	1964	**9**
1954	**1**	1965	**16**
1955	**2**	1966	**16**
Total	**97**		

*Source: *Radiation Exposure of Uranium Miners*, Hearings before the Subcommittee on Research, Development, and Radiation of the Joint Committee on Atomic Energy, U.S. Congress, 19th Cong., 1st sess. Information supplied by AEC, as obtained from PHS, Salt Lake City Field Station. Note: 1 of 97 deaths is not associated with a particular state.

they had not. He had taken the high ground in this political skirmish and refused to give it up. He opened up his testimony with a powerful verbal volley:

> After 17 years of debate and discussion regarding the respective private, State and Federal responsibilities for conditions in the uranium mines, there are today—or were when these hearings were called—no adequate and effective health and safety standard[s] [or] inspection procedures for uranium mining.
>
> There is unmistakable evidence of a high incidence of lung cancer among uranium miners; 98 have died from it, and another 250 to 1,000—the estimates vary—are already incurably afflicted with it.

> The best available evidence is that over two-thirds of the approximately 2,500 underground uranium miners are working under conditions which at least triple their prospects of dying from lung cancer if they continue this work and these conditions remain unchanged.

Wirtz made it clear that the Public Contracts Act gave him the power to clean up the uranium mines and to impose standards. Wirtz said he was forced to do this because no one else would. The studies of the problem by government agencies were deplorable because none of them caused anyone to act. "It is a record, nevertheless, of literally hundreds of efforts, studies, meetings, conferences, and telephone calls—each of them leading only to another—most of them containing a sufficient reason for not doing anything then—but adding up over a period of years to [a] totally unjustifiable 'lack of needed consummative action.'"[34]

Wirtz's standard was immediately attacked as being arbitrary and unworkable, a standard that the mining industry could not meet. Wirtz did not flinch. He thanked the committee and left. The next day, Wirtz's position was strengthened by the testimony of Leo Gehrig, the acting surgeon general and head of the Public Health Service. Gehrig painted a dismal picture and made dire predictions:

> Our projections indicate that in the group that has already been exposed there is a possibility of a total of 529 deaths due to lung cancer. The additional deaths in the already exposed group will occur over a period of about 20 years.
>
> Little can be done to prevent deaths among those who have received an excessive dose. However, it is imperative that an immediate action program be taken to reduce the radiation exposures of the uranium miners. The Department of Health, Education and Welfare is recommending that this control program be based on one which limits the annual exposures of the uranium miners to not more than 4 working level months (WLM).[35]

Gehrig was proposing a standard very close to what Wirtz had already ordered, pending his thirty-day comment period. The "working level" term had been modified to include a "working level month." This was the amount of radiation a miner was exposed to in a month, or 168 hours (42 hours per week times four weeks). A mine that had 1 working level would give a miner 1 working level month after a month of work in the mine. A mine containing 10 working levels of radiation would give a miner an exposure of 10 working level months in just one month. Over the course of a year, the same miner would get an accumulated dose of 120 working level months of exposure. Wirtz's proposed 0.3 working level would limit a miner to a total of 3.6 working level months after a year of work. Gehrig was suggesting the limit be 4 working levels per year, or a 0.33 working level per month.

Gehrig explained that after seventeen years, the study on the uranium miners had developed statistical proof that increased exposure to radiation created increased risk of death by cancer and that exposure had a cumulative effect:

Estimated Number of Miners with More Than 120 WLM of Cumulative Exposure to Radon Daughters as of Dec. 31, 1963*

CUMULATIVE WORKING LEVEL MONTHS:	NUMBER OF MINERS IN CATEGORY
Less than 120	961
120 to 359	920
360 to 839	739
840 to 1,799	482
1,800 to 3,719	246
3,720 and over	67

Total 3,415

*Source: *Radiation Exposure of Uranium Miners,* Hearings before the Subcommittee on Research, Development, and Radiation of the Joint Committee on Atomic Energy, U.S. Congress, 19th Cong., 1st sess. Submitted by the deputy surgeon general, Public Health Service, for the record, on July 5, 1967.

Among those workers in the study group with cumulative exposures less than 120 (working level months) WLM, the age-adjusted lung cancer incidence rate is not significantly greater than for the comparison group.

Among those miners with exposure between 120 and 840 WLM, the observed incidence rate is 4.9 per 10,000 man years as compared with an expected rate of 3.1.

In the group with exposure between 840 and 1,799 WLM the observed incidence rate is 18.2 compared to the expected rate of 3.9.

In the group with exposure between 1,800 and 3,719 WLM the observed incidence rate was 50, as compared to an expected rate of 4.8.

Among those with cumulative exposures in excess of 3,720 WLM, the observed incidence rate was 171, as compared to an expected rate of 4.3.

For the latter three groups the increased rates, of roughly 4.7, 10, and 40 times the expected incidence rate, support the interpretation that there is a causal relationship between excessive exposure to radiation, as encountered in uranium mines, and an increased incidence of lung cancer.[36]

Gehrig said that with the present conditions at the mines, where the average readings across the Colorado Plateau were a constant 2.1 working levels, 246 more people would die of lung cancer over the next twenty years than would die if they had never set foot in a mine. If the limit of 1 working level was enforced in the mines, only about 96 more deaths than expected would be found. However, if the level was reduced to a 0.3 working level, as Wirtz had ordered, the expected lung cancer deaths would fall within the national average. In other words, at a 0.3 working level, the uranium miners would be exposed to no more danger than anyone else.[37]

The opposition began to fight back with a variety of bizarre arguments based on untenable positions. One of the most aggressive opponents of imposing standards on the uranium mines was Representative Holifield, who was immediately defensive because the AEC and his committee had come under attack from Wirtz and the press for remaining idle while uranium miners died and were dying. As Gehrig concluded his testimony, Holifield suggested that studies existed which proved that mice actually develop a tolerance for radiation, just as they do for poisons such as strychnine. Humans probably do the same, he said. A study at the Oak Ridge, Tennessee, laboratory of the AEC seemed to indicate that an animal could survive multiple low doses but not one large dose of radiation. "In other words, nature took care of the absorption and rehabilitative effect on the mouse's physical body when they [*sic*] received it [radiation] in small doses. It is analogous to a heart patient taking strychnine. He takes little doses over a series of hours, days or months, when if he took the cumulative amount at one time it would kill him. But his body develops a certain tolerance...on small doses over a long period of time."

Gehrig politely engaged in this exchange with Holifield, who suggested that the conclusions drawn from seventeen years of experience with human uranium miners, not mice, was still nothing but theory. "I have no objection to your believing what you want to believe. Do you believe there is a straight-line type of relationship between exposure and cancer incidence?" Holifield asked.

"We have postulated in our hypothetical model that this is true," Gehrig repeated.[38]

If there was any weakness in the medical theory, it was that most of the miners smoked cigarettes. Some of the miners who died of lung cancer had been heavy smokers. Holifield zeroed in on this fact. He threw back at Gehrig a number of cases in which miners had worked less than a year in underground uranium mines but had smoked cigarettes for thirty-six to forty years. "You tell me if I am to believe statistics based on that kind of input at that low level, at that cumulative input. For 12 months he worked underground. He died at age 71. He smoked 2 1/2 packs of cigarettes for 60 years, and he is listed as a cancer death by radiation."[39]

Gehrig quickly defended himself and the conclusions that were drawn from the study. He admitted that cigarette smoking was a complicating factor but did not discount the studies. "We are unable to differentiate between a malignancy occurring from radiation with all definiteness in an individual, and that occurring in cigarette smoking." Gehrig noted that the average age of the lung cancer victims was fifty-five, which meant that an equal number were under that age and were in their forties—some even younger. Lung cancer under the age of fifty-five was a "rarity," he said. Also, more than half of the lung cancer types were oat cell cancers, he explained. The occurrence of this type of lung cancer among nonuranium miners was only 17 to 20 percent. "The factor of age of onset, the factor of type of tumor, gives not only the clarity of understanding for a given case, but points to another thing in the chain of evidence, that there is a relationship between cancer and radiation."[40]

In an attempt to regain its dominance, the AEC took the offensive, countering with its own set of recommendations to control conditions in the mines; these were, of course, less stringent than those proposed by Wirtz. The AEC suggested that 1 working level month be the general standard, but a miner could exceed that exposure and accumulate up to 6 WLMs in just three months. This would allow a miner to work for up to three months in a mine that registered 2 working levels of radiation, or 600 picocuries. Or, the same miner could work for one month in a mine that had 6 working levels, or a constant exposure rate of 1,800 picocuries. The total annual dose would be limited to 12 working level months. The AEC suggested that efforts be made to reduce the radiation levels as much as possible but that mining companies should be given two years, or until July 1, 1969, to comply with the regulations.[41]

This proposal went directly counter to Wirtz's proposal. Wirtz had stated that the urgency of the situation was clear and required immediate action. The AEC was essentially saying that nothing should be enforced for another two years, enough time for some miners to be exposed to deadly levels of radiation.

The AEC also came up with an "action plan." The plan was well thought out and covered almost all aspects of the radiation problems in uranium mines. It assigned areas of enforcement and monitoring to a variety of federal agencies, including the interior department's Bureau of Mines, the AEC itself, the Public Health Service, and the Department of Labor. It covered everything from improving ventilation and sealing mine walls to filtering mine air in order to reduce radon in the mines. Monitoring was to be improved with the use of dosimeters, or individual radiation badges, so that accurate cumulative doses for each miner could be determined using state-of-the-art technology. Respirators were to be used, further research was to be conducted, and the agency would work to improve workmen's compensation. It was an impressive program. However, it remained on paper only and was never implemented.[42]

Industry quickly began to mount its opposition. Dr. Robley Evans, the Massachusetts Institute of Technology radiation expert who had set the initial exposure standard for people working around radiation at 10 picocuries, was brought before the committee to testify on behalf of Kerr-McGee Oil Industries, which had hired him as a consultant in 1961. Evans was eminent in his field, and his testimony was a wonderful public relations move for the sprawling oil company turned uranium miner.

Not unexpectedly, Evans politely recommended that Wirtz's proposed standard of a 0.3 working level be tossed out. Singing a corporate tune, Evans said that an "orderly" progression from existing levels of exposure to what was proposed by the AEC should take place: "For example, a sequence of 30 WLM (working level months), followed by 24 WLM for the second year, 18 WLM for the third year, and 12 WLM per year thereafter would be commensurate with such an orderly transition to an annual average of 1 WL, or 12 WLM per year."[43]

Such a proposal would allow a miner to continue working in mines with readings of 3 working levels, or more than the average of all the mines on the Colorado Plateau. Evans's assault on Wirtz's proposal did not end there. Opposing Wirtz's viewpoint, Evans suggested that the current studies were inadequate. "It is also clear that the time is ripe for an enlarged and detailed retrospective and prospective epidemiological study of the American underground uranium miners. Considering the large turnover in the labor force it is probable that on the order of 20,000 miners have been involved to date. It is just as important to locate and study those who had minimal exposures as it is to study the high-level cases."[44]

The need for federal control over the problem was never more apparent than when the mine inspector for the state of New Mexico testified. The state had a fairly aggressive mine control program compared with other uranium mining states such as Arizona, which had very little enforcement. Don B. Buddecke, the deputy state mine inspector for New Mexico, said the state worked with uranium companies to get the radon readings down to 1 working level. Their enforcement program was to close a mine down if it had a reading of 5 working levels and demand immediate correction of conditions if the reading was at 3 working levels. Buddecke said that because the medical testimony was inconclusive, lung cancer had not yet been "validated or accepted" and that the proposed standards were excessive. As if he were working for a uranium company, Buddecke said that the proposed standard would cripple the uranium mining industry if it were enforced.

He even raised the national security issue: "After careful consideration of the proposed new radon level regulations of essentially a 0.3 working level, it is our viewpoint that this does not constitute a realistic level or one that will enable the underground uranium mines to produce and function as they must and that a strict enforcement of this standard could seriously impair the uranium production in the State of New Mexico with possible serious results and effects in the country."[45]

In the midst of this immense display of concern for the welfare of the mining corporations, Buddecke, as a public employee, failed to speak on behalf of the thousands of men working in the mines whose lives were on the line. With testimony like this, mining companies had no need for their own expert witnesses.

This state official's testimony was mild compared with that of the corporate representatives themselves. Mining company officials testified that reducing radon and radiation in the mines was actually hazardous to the miners' health! "We believe that the means which would have to be followed to comply with that [Wirtz's] directive would unnecessarily greatly increase the hazards to the uranium miners employed by Kerr-McGee," said M. F. Bolton, vice president of minerals for Kerr-McGee.[46] Bolton, in lengthy and detailed testimony, attempted to convince the committee that it was simply too difficult to reduce the radon hazard to the level that Wirtz was calling for.

Kerr-McGee entered the uranium mining business on the Navajo Reservation in 1952, operating in the Cove and Mesa mines in the Red Valley area west of Shiprock. They initially employed about 50 Navajo miners and by 1963 had more than 150 Navajos on their payrolls, Bolton explained. "During this 11-year period, frequent inspections were made by the U.S. Bureau of Mines, state mine inspectors, and Navajo Tribe inspectors. Working conditions were maintained at a high level and programs for improvement were continuously and vigorously prosecuted," he said. Under further questioning about the use of Navajos in the mines, Bolton suggested that the company was doing a favor to the Navajos by hiring them to work in the mines. Bolton also suggested that the uranium belonged to his company and not the Navajos.

Sen. John Pastore of Rhode Island, who chaired the JCAE, pushed a line of questioning about the use of Navajos:

> Pastore: Can you explain why you have the Navajo Indians? Is it because other people are not attracted to this kind of occupation?
>
> Mr. Bolton: No sir. When we first discovered our uranium in 1952 on the Navajo Indian Reservation, part of the understanding that we had with the Navajo Tribe was that we would employ Navajo Indians in as many occupations and use them as much as we could. We found that they were quite proficient, made very good miners, if trained, and therefore we just chose to use 100 percent Navajo miners on our Navajo Reservation properties.
>
> Pastore: In other words, it was an accommodation to them?
>
> Bolton: Yes, sir.[47]

It was not surprising, then, that Bolton concluded his testimony by predict-

SEN. JOHN PASTORE OF RHODE ISLAND DEFENDED MINING PRACTICES ON THE NAVAJO RESERVATION.
PHOTO: AP/WIDE WORLD PHOTOS

ing the demise of the entire industry: "Unattainable, unrealistic standards such as those proposed by Secretary Wirtz would shut down virtually all presently operating underground uranium mines, would largely remove the incentive for exploration, and would seriously impair the domestic uranium industry's capability to supply the uranium required by the rapidly expanding nuclear power business."[48]

While company officials were given almost unquestioned latitude to say what they wanted, union representatives rarely went unchallenged. Although uranium miners were never unionized for unknown reasons other than the remoteness of the mines, the Oil, Chemical and Atomic Workers International was given time in front of the committee. Anthony Mazzocchi, legislative director and lobbyist for the union, pointed out some clear examples of discrimination against the Navajo uranium miners. While the medical studies conducted by the Public Health Service followed the usual procedure of separating the data by race, the medical evidence presented to the committee was for white miners only. Mazzocchi pointed this out in his blunt and emotional testimony:

> In the past, we have been told to go to the States. We know and we told this committee the states would be ineffective. The standard which has been adopted comes late. No one denies, no matter what the conflict about statistics, that there are dead miners who are dead today because they inhaled radon gas on the job.
>
> How many dead miners justify the lack of ventilating equipment? How many families must be broken before this industry takes steps to eliminate the hazard? I for one think the figures which have been presented before this committee are incomplete.

Shakespeare had Shylock say:

"If you prick us, do we not bleed? If you tickle us, do we not laugh? If you poison us, do we not die?"

I say if a Navajo Indian breathes radon gas in the mines, he is injured just as much as a white man. It seems that the information we have is incomplete.

We are suffering through a repetition of the Triangle Shirtwaist fire on March 25, 1911. The number of dead bodies are probably not identical, but the moral issues are identical. Do we establish industry solely for the benefit of those who would make a profit and then set a safety standard compromised on economic considerations, abolish the long-term program of "safety first," and seek to berate those who take a belatedly feeble step to end the tragic loss, the unnecessary loss of human lives so that some bookkeeper can add a few more pennies to the profit side of the ledger?[49]

Langan W. Swent, the vice president of the Homestake Mining Company, the Homestake-Sapin Partners, and United Nuclear, one of the largest uranium mining operations in the West, underscored Bolton's earlier testimony by saying that compliance with the 0.3 working level standard would require so much air to be moved through the mines that more problems would be created than solved: "Serious eye injuries could result from the larger sand particles the air would carry. Bronchitis and pneumonia incidence would greatly increase. Verbal communication would be very difficult causing misunderstands [*sic*] which could lead to accidents."[50]

The lengths to which the mining companies would go to keep from being forced to spend the money needed to make the mines safe for humans was typified by Richard D. Bokum II, the president of Bokum Corporation, which had acquired United Nuclear Corporation and accounted for about 25 percent of all the domestic uranium being produced at the time. Bokum attempted to discount the validity of the medical studies by saying the results were inconclusive and actually proved that uranium mines were safe. He specifically mentioned a casual study of miners in the Grants, New Mexico, and Ambrosia Lake district which showed no elevated cancer levels to date. "Actually, Mr. Chairman, I could make a case in looking at some of the figures in this Report No. 8—I could make a case, if I wanted to be facetious, where the miners' health was improved by working underground and being subject to radon daughter products. There are some areas in this chart where it shows that the expected number of cases of lung cancer should be so many and the people working in the mines have zero cases."[51]

Determined to fight the proposed standard, the major mining companies in the uranium industry banded together to form the Committee on Mining and Milling, Atomic Industrial Forum, Inc., and issued a joint statement. It was signed by twenty-three uranium mining companies, including Kerr-McGee, Homestake Mining, Anaconda, Union Carbide, and Vanadium Corporation of America. It predicted that, "A grave situation would result if the 0.3 WL is adopted as the maximum exposure level." The consequences were predicted to be the closure of at least two-thirds of the underground uranium mines; and this halt in the flow of domestic uranium would mean the end of new nuclear power plants and the possible loss of this major source of electrical power for the country.[52]

Throughout this thrashing and wailing, Wirtz remained unswerving in his commitment to make the uranium mines safe. On June 10, 1967, after about a month of testimony and reaction, Wirtz issued his revised regulation. Only minor changes were made, including an eighteen-month transition period for the industry to comply. In customarily tough language, he blasted the critics of his standard, who were using the argument that because there was no unanimous agreement among the scientific community, no action should be taken. He argued that because there was doubt, and lots of it, that alone was cause for strict standards:

> Taken together, these statements reflect what has prevailed as the operative enforcement, or non-enforcement, principle for twenty years: that disagreement among scientific authorities about just how much radiation exposure a man's lungs can take, and difficulty in measuring this exposure precisely, justifies not having a standard and not enforcing the law.
>
> That operative principle gave death an unconscionable advantage—which it took.
>
> The operative principle underlying the Regulation issued today is that doubt—the disagreement of experts—is reason for not taking a risk instead of excuse for taking it—when the stakes are cancer. Life, not death, is entitled to the benefit of the expert's reasonable and responsible doubts.[53]

Wirtz concluded his statement by threatening to close down any mines that did not comply with the standards within the amount of time they were given.

Finally, on August 8, three months after Wirtz had made his stand and after a period of review and comment, mostly from angry politicians and industrialists, the 0.3 working level would become the standard as of January 1, 1969. As an intermediate step, mine operators were required to immediately reduce the mine hazard to 1 working level or shut down until they did. In his final testimony before the committee in 1967, Wirtz outlined the compliance situation. None of the disasters the mining companies had predicted had occurred.

Wirtz said that steps were being taken to enforce the Walsh-Healey Public Contracts Act, and that since the initial order was issued, nearly 70 percent of all the underground uranium mines had been inspected. Of those, twenty-three had readings under a 0.3 working level, twenty-three had readings under 1 working level, thirty-nine had readings between 1 and 3 working levels, and forty-one had readings above 3 working levels but under 8 working levels. Of the remaining mines, some had readings of over 30 working levels. Wirtz also reported that the majority of mine operators were complying with the regulation and were not going out of business. Of eighty-one responses to his order, only one mine operator said it was impossible to meet the standard, and nine others refused to cooperate. Only two mines decided to close down, and the rest agreed to meet the regulations.[54]

REP. WAYNE ASPINALL OF COLORADO SPOKE ON BEHALF OF URANIUM MINING COMPANY PRACTICES ON THE COLORADO PLATEAU.
PHOTO: AP/WIDE WORLD PHOTOS

Wirtz had won. A tough regulation had been set that would ensure some safety not only for the uranium miners on the Navajo Reservation, but for all uranium miners in the United States. The economic crisis that had been predicted by both large and small mine operators had not materialized. Mine ventilation was a well-developed technology, but because the mining companies did not want to give up a percentage point or two of their profits, no action was taken. Until Wirtz, no one in the federal government had taken the initiative to curb the well-known and well-documented dangers.

Even this long-overdue effort, however, did not achieve its objective, for the standard was not widely adopted by government agencies. However, the Federal Radiation Council continued to hold hearings on exposure standards and ultimately agreed on a standard of 1 working level in the mines. This translated into workers being exposed to 12 working level months of radiation over the course of a year.

Wirtz's regulations came exactly seventeen years after Holaday had started the uranium miners' health study. This was also the average time it had taken for European pitchblende miners to succumb in large numbers to lung cancer. Just as Holaday and others had feared, uranium miners, including the Navajo miners, were dying of cancer seventeen years after beginning to work in the mines. The regulations had come too late. There was nothing left to do but wait and watch as increasing numbers of miners fell. The death watch would continue for another twenty-five years.

In an almost conciliatory, apologetic tone, Congressman Wayne Aspinall of Colorado told Wirtz in the final hearing that all the mining companies wanted to do was to serve their country. Aspinall explained why mine operators had been generally cooperative despite their objections to the regulations: "If you represented these people as I do, you would expect this because this has been the history of all time. When they were given the call to serve their country by furnishing uranium, they did what was expected of them."[55]

Aspinall was speaking for a dozen or so mining companies who produced uranium at a handsome profit. He somehow forgot about the thousands—estimates were that at least 10,000 or more people worked in the mines and mills—of men and women exposed to dangerous levels of radioactivity. These people, too, had responded to the call to serve their country and worked in the mines and mills that produced the uranium that went to the nation's defense. These people gave much more than the mining companies ever could. The mining companies were being asked to give up a little of their profits to remove the radiation danger they knew was present. The miners and millers had given and were giving their health and their lives. They were never told what their loyalty would cost them and their families.

7 The Fight for Justice

IN 1967, NAVAJO MINER PETER YAZZIE had a gnawing pain in his left side. It wouldn't go away. He lived in the community of Red Rock in Arizona,* and had worked in the uranium mines in the Cove mesa area since 1950 when companies like the Vanadium Corporation of America, Kerr-McGee, and U.S. Vanadium had begun digging and advertised for workers. After seventeen years of hard, physical labor in the uranium mines that had been nicknamed "dog holes," Yazzie went to the Indian Health Service hospital in Shiprock. Doctors told him to have a chest X-ray. The results were frightening. He had cancer in both lungs. The disease was so far advanced that the doctors could do nothing. Yazzie was given pills to take when the pain became too bad. Three years later, at the age of forty, Peter Yazzie knew the end was near and was driven to a hospital in Albuquerque. He died eight days later on June 6, 1970. He left a home that was a simple adobe hogan heated with wood; a wife, Dolores, age thirty-six; and ten children, ranging in age from two to eighteen. His wife began to collect $250 a month on which to raise her family. When she told a newspaper reporter about her husband's death some years later, she sat hugging his photograph while tears streamed down her face.

---

* Red Rock officially changed its name to Red Valley in the late 1970s.

In the nearby Red Valley community of Oakspring, another Navajo uranium miner named Clifford Yazzie (who was not related to Peter Yazzie) complained that he always felt ill. Skin problems prompted him to make a trip to the Shiprock Indian Health Service hospital. Tests showed that his body was riddled with cancer. It was in his lungs and intestines. Ultimately his spinal cord was affected, and he became paralyzed from the waist down. He died in April 1972, leaving his wife Fannie and fourteen children. The family survived on small cash payments from Social Security and a small flock of sheep. Mrs. Yazzie pawned some treasured family jewelry to buy groceries and feared that she would never be able to retrieve the valuable silver and turquoise.

Todacheenie Benally was still working in the uranium mines when he began to suffer from a severe pain in the left side of his chest. Chest X-rays showed advanced lung cancer. He quit work to spend his remaining time with his family. He died on May 28, 1972. Todacheenie died before his last child, Alfreda, was born, and left a wife, Rose, and thirteen children who survived on Social Security payments and some sheep. Rose Benally told a reporter a year after her husband's death she would like $100,000 to provide a good home and education for her children. It was something that she could only dream about.

There were others. Billy Johnson, a uranium miner for twenty years, became sick while working in the mines. He died of cancer on May 21, 1972, leaving his wife Louise and five children. Johnson's death was preceded by the death of a friend, Lee N. John, also a longtime uranium miner, who died on January 19, 1972, leaving his wife Mae and eight children, including twins. Both women tended small flocks of sheep, hauled water, and chopped wood for cooking and heating. Mae John's eldest daughter, Mary, passed up a chance for higher education to take a job as a cook at the Cove Day School to help her family survive.[1]

There were other uranium miners who had died beginning as early as 1963 and 1964, but the handful of deaths and increasingly serious health problems in the Red Valley community could not be ignored. The deaths and illness were troubling to local tribal leader Harry Tome. At the time Tome was a council delegate, a man elected from his community to the Navajo Tribal Council, the highest governing body in the Navajo Nation. Along with Elwood Tsosie, the president of the Cove Chapter, the local tribal unit of government, he decided they needed help with what was clearly a serious problem.

What Tome and Tsosie were witnessing was what the Public Health Service, the Atomic Energy Commission, federal officials, and members of Congress had known for years. Nevertheless, despite regular health exams, despite dozens of studies, and despite a series of congressional hearings during the summer of 1967 and widespread media coverage, no one had bothered to inform and warn the Navajo miners of the dangers the mines held. Navajo uranium miners across the reservation were dying, but they did not know why.

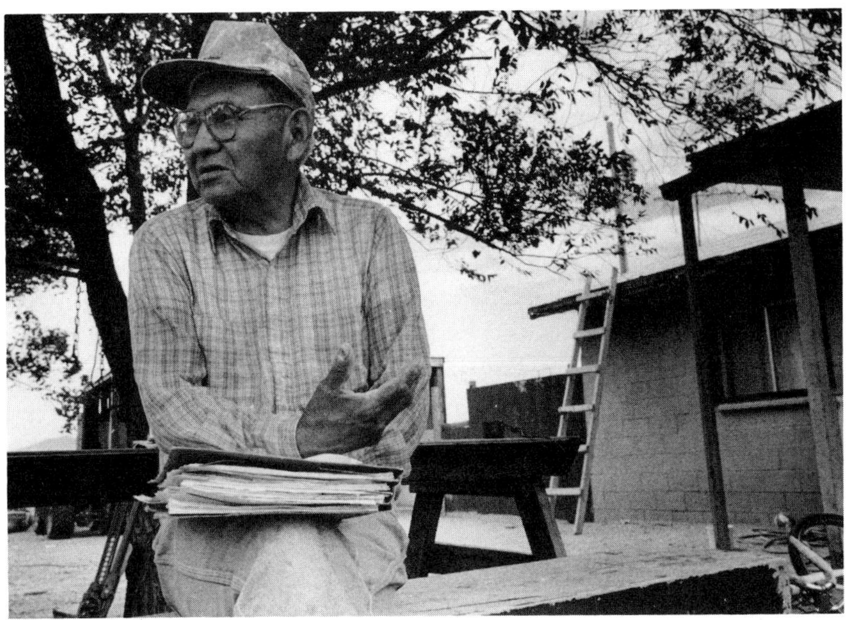

NAVAJO HARRY TOME AT HIS HOME IN RED VALLEY, ARIZONA. TOME WAS INSTRUMENTAL IN BRINGING CONGRESSIONAL ATTENTION TO THE PLIGHT OF NAVAJO URANIUM MINERS.
PHOTO: MURRAE HAYNES

In 1963, Tome went to work for the minerals department of the Navajo Nation. Interviewed in February 1993, he said, "I visited a lot of these [uranium] mines. I heard complaints back then of health. It never occurred to me it could be because of the uranium." The complaints of poor health among the uranium miners persisted. As a longtime member of the Native American Church, he heard more complaints of lung problems from his fellow worshippers at the regular prayer meetings he attended. "Nine times out of ten they worked in the uranium mines," Tome said. "The people who ran the services said it had something to do with the uranium." Tome did not at the time understand how it could be. "I began to make some inquiries and talked to doctors. They said, 'yes,' there are studies that show it is harmful." The concept of radiation was foreign to most Navajos at the time, Tome said. The doctors described it to him as a "dangerous smoke you breathe in." Tome, educated as an engineer, understood clearly but knew he had a difficult job on his hands. "Slowly we began to educate our leaders," he said.

He first stopped at a Red Valley chapter meeting and advised the group to get some help. Far too many members of the chapter were dying. They needed help. Tome told the chapter members what he had been told by the doctors.

They were upset. None of the miners had ever been warned of the dangers. Who was at fault? Why hadn't either the mining companies or the government told them? Why hadn't the mines been made safe? "That was the beginning," Tome recalled. The chapter council needed some convincing but finally drafted a resolution seeking help from the tribal council in Window Rock, Arizona. The resolution stated in effect that "we should be compensated. We feel we have been mistreated. We were never told of the dangers," Tome said.

There was very little the Navajo council could do about the problem, Tome was told. The tribe had no excess funds, yet they knew the miners and their survivors needed to be compensated. The help and the compensation should come from the federal government. The word had to be spread, Tome realized, and the news media was the quickest way to reach congressional representatives in New Mexico and Arizona. A friend knew of a reporter at the *Albuquerque Tribune* who might be willing to write a story about the miners.

On August 17, 1973, a story by staff writer Urith Lucas appeared across the top of the front page under the headline "Navajos Who Mined Uranium Dying from Lung Cancer." Accompanied by Tome and Tsosie, who acted as translators, Lucas interviewed more than a dozen suffering miners and widows of miners who had died of lung cancer. The story was given large play. The extent of the problem was unmistakable. Clearly something had to be done.

As part of the story, Lucas contacted two members of the state's congressional delegation, the late Sen. Joseph M. Montoya and Rep. Manuel Lujan. Both promised to help by sponsoring legislation, and Montoya agreed to help the Navajo widows with their claims for Social Security benefits. Lujan, who at the time was a member of the Joint Committee on Atomic Energy, was cautious and, echoing statements made by other members of the JCAE, said, "this is a very complex problem and cannot be solved by legislation alone."[2]

The *Tribune* article, coupled with personal lobbying by Tome, resulted in the introduction of legislation. In October 1973, Senator Montoya announced that a bill would be introduced into Congress to provide from $169.80 a month for a single survivor to up to $339.50 for a disabled miner with three dependents. Montoya estimated the bill would affect up to 1,000 disabled miners and their survivors. The bill required a miner to have worked in the mines for at least six months and to have contracted cancer at least five years after beginning work. It also prevented the government from rejecting claims unless the government, not the miner, proved that the cancer was not caused by radiation exposure.[3]

Montoya's bill was introduced as an amendment to Senate Bill 1029, which had been introduced earlier in 1973 but which was related to government compensation to coal miners suffering from black lung disease. Montoya's amendment extended the benefits to cover Navajo uranium miners. Montoya said that an estimated 15,000 to 20,000 men and women had worked in the ura-

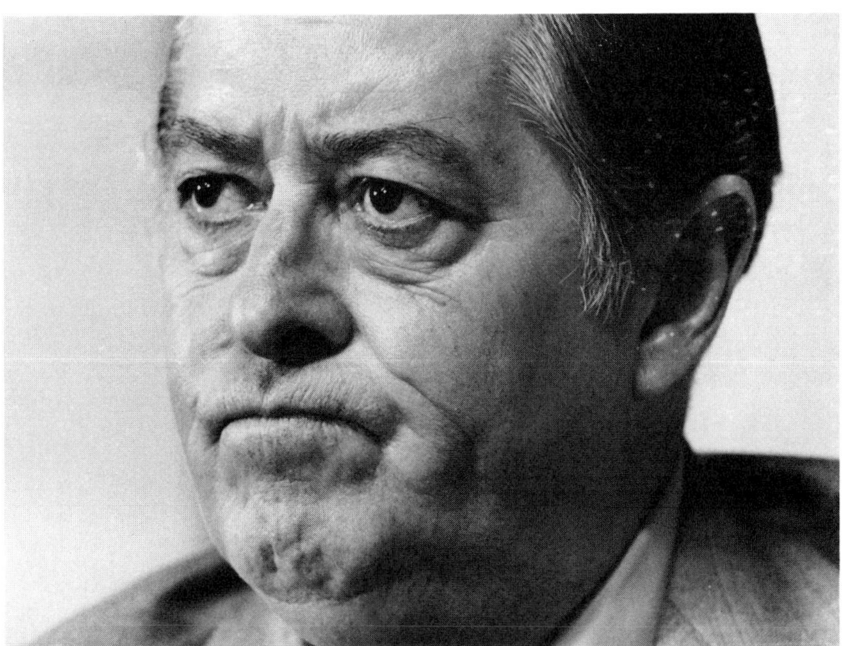

SEN. JOSEPH M. MONTOYA OF NEW MEXICO INTRODUCED LEGISLATION IN THE EARLY 1970s TO COMPENSATE STRICKEN NAVAJO MINERS.
PHOTO: AP/WIDE WORLD PHOTOS

nium industry on the Colorado Plateau and that from 1,000 to 2,000 of them would most likely die of lung cancer.[4] About two months later, shortly before Congress recessed for the Christmas holidays, Lujan introduced a similar bill, H.R.11567, which duplicated the black lung disease bill being considered in the Senate. Lujan also included uranium miners in the bill.

Six months later, with the bills stalled in both houses, Bureau of Indian Affairs Commissioner Morris Thompson announced that his agency was looking into ways it could provide assistance. He also said that the mining leases that were issued to companies such as Kerr-McGee may have placed responsibility on the Bureau of Mines to ensure mine safety. He contended that the Bureau of Indian Affairs was not responsible for mine safety on Indian lands.[5] This lack of result would characterize compensation efforts for the next twenty years.

The next five years were ones of deep frustration for Tome. Repeated letters and repeated trips to Washington to testify on behalf of the afflicted Navajo miners resulted in nothing. In 1974, two bills, H.R. 8107 and Senate Bill 1028,

REP. MANUEL LUJAN, LEFT, AND SEN. PETE DOMENICI, BOTH OF NEW MEXICO, WERE ALSO ACTIVE BUT UNSUCCESSFUL IN OBTAINING COMPENSATION FOR NAVAJO URANIUM MINERS.
PHOTO: AP/WIDE WORLD PHOTOS

were introduced and contained the same requests for compensation that the previous bills had contained. Rep. Harold Runnels, who represented the Gallup area and most of southern New Mexico, and who had become involved in the issue, introduced the bill on the House side. On May 24, 1975, Runnels held a public hearing on Indian health issues in Gallup and told the assembled crowd that he would reintroduce H.R. 8107 of the previous year.[6] That same summer, the Shiprock Agency, a middle level of government in the Navajo tribe, passed a resolution urging the U.S. Senate to pass Senate Bill 2018. As president of the Shiprock Agency Council, Tome had urged the council to launch the request. It was approved 48 to 0. Their requests fell on deaf ears.

However, Tome refused to quit. The next year, 1976, both bills having died, the Red Valley Chapter again launched a move to get some congressional action by passing a resolution that not only asked for assistance but for an organized study of the afflicted Navajo miners. The resolution asked that all congressional representatives in New Mexico and Arizona reintroduce bills similar to the ones that had died. In addition, the chapter asked the director of the Navajo Area Indian Health Service to establish a special clinic with the capability to screen for lung carcinomas at each of the five agencies and to interview and examine all those who may have been exposed to unsafe radiation levels.[7] The resolution went on to ask that every chapter on the Navajo Reservation develop a list of miners and surviving family members.

Time passed. Miners continued to die, leaving poverty-stricken widows and children. A year later, in 1977, still nothing had been done. Tome persisted. He wrote again to congressional representatives, asking that bills again be introduced to provide compensation to afflicted miners. Representative Lujan responded by expressing his concern and saying that he had introduced H.R. 5529 to provide compensation.[8] New Mexico's Sen. Pete Domenici wrote to Tome on March 25, 1977, that Senate Bill 717, which promoted mining safety, was scheduled for hearings on the next few days—March 30, 31, and April 1—but did not contain a specific provision to compensate dependents of deceased or disabled uranium miners.[9] Domenici said he was working on amending the bill to include this compensation.

Tome quickly made plane reservations and flew to Washington in hopes of getting something for the uranium miners included in the bill and passed. Tome went by Representative Runnels's office but was unable to speak with him. Tome flew home, once again empty-handed. Runnels wrote to Tome and said he was working on updating the Respiratory Disease Benefit Act (the bill Tome wanted reintroduced) and would introduce a new bill into the House as soon as the updating was complete.[10] There was little that Tome could do other than wait and hope.

It was now the middle of 1978. After five years of trips, pleadings, and promises, Tome had gotten nowhere. Twenty-five miners in the Red Valley Chapter, a community of several hundred families, had died of lung cancer. The story was being repeated in other parts of the Navajo Reservation. Tome realized two things. First, he needed some powerful help. Going it alone was getting him nowhere. Second, appealing to Congress was also getting him nowhere. He had to take a different path, and this meant going to court. Representative Runnels suggested that Tome contact Stewart L. Udall.

Udall had grown up near the Navajo Reservation and came from a politically powerful family. He had been a member of Congress from Arizona since the mid-1950s. In 1961, Udall was appointed secretary of the interior and stayed on through the presidency of Lyndon Johnson, until 1969. He was living in Washington at the time and working out of the law offices of Duncan, Brown, Weinberg and Palmer on Pennsylvania Avenue, not far from the White House.

Tome mulled over Runnels's advice. It seemed like the right thing to do, and Udall seemed like the right man. He knew about and understood Navajos. He was well known and politically well connected. Udall knew the legal and congressional processes. If anybody could get the job done, Udall could. Still, Tome was hesitant. He had been working on this problem for five years. He had made so many trips to Washington, D.C., that he could not remember them all. Congress had listened but had done nothing. Why should he expect anything to change? The white man's world was a cold and forbidding place. Once again Navajos were dying at the hands of the white man, just as his ancestors had 100 years earlier. Would it ever change? Tome stood on the sidewalk at 1775 Pennsylvania Avenue outside the tall building where Udall worked, and watched the glass doors open and close as people rushed in and out. He was nearly two thousand miles from the peaceful mountains near his home in Red Valley. He thought of the blue sky, the sun, wind, and rain. He thought of his friends and neighbors. They were dying. He took a deep breath and went in. "I was tempted to go home and forget about it," Tome recalled. I was just a little guy and a Navajo. But my people needed help."

On the seventh floor, Udall was surprised when he heard that a Navajo named Harry Tome was waiting to see him. Udall listened to his story and was saddened by what he heard, both about the plight of the Navajo uranium miners and Tome's frustrating quest to obtain compensation for the victims. Udall knew he had an issue he could sink his teeth into, but it would take time and money. Udall told Tome he would like to help. Tome had in fact come to the right man because Udall was already at work on a lawsuit involving claims by victims of the atmospheric testing of nuclear weapons in Nevada in the 1950s—the so-called "downwinders," who lived in southern Utah and northern Arizona. That spring, Udall and a cousin had taken a quick automobile tour of the area to see the damage for themselves. A suit involving devastated uranium miners was part of the same overall problem, Udall realized.

Public service was not new to him. The son of a state supreme court judge, Udall had grown up in the small town of St. Johns on the eastern edge of Arizona. As secretary of the interior, he was instrumental in getting a host of environmental bills into law which have had a lasting effect on this country. They include the Endangered Species Act; the Wilderness Act of 1964, which placed millions of acres of public land under permanent protection; and the Wild and Scenic Rivers Act of 1968. In addition, he established four new national parks, fifty-six federal wildlife refuges, eight national seashores, nine national recreation areas, and twenty-two national historic sites.

On the way back home in 1978, Tome felt that finally something was about to happen. He told his friend Elwood Tsosie about his meeting with Udall, and the two composed a letter to Udall, making their plea for his help formal. "Hundreds of Navajo miners worked in these uranium mines in Arizona, New Mexico, and Colorado. During these mining operations, Navajo miners were exposed to high levels of radiation and consequently twenty-five (25) miners from our community died and many others are suffering from avaplastic [*sic*] carcinoma of the lung.

"Since you are aware of this problem and would like to help us, we are seeking your support to overcome this unfortunate circumstance."[11]

Udall was already developing a legal war on two fronts. First, he teamed up with lawyers battling for the right to compensation for downwinders in southern Utah. Second, he contacted his old friend, a former ambassador under presidents Kennedy and Johnson, Phoenix attorney Bill Mahoney, to help with the uranium miners' case. The two developed a plan of attack for their suit and contacted the Navajo general counsel, George Vlassis, to ensure that any potential lawsuits would not conflict with anything Vlassis or the tribe might have planned. Udall and Mahoney thought about going directly to Congress but dismissed the idea. Tome's experience with Congress greatly influenced their decision. Bill after bill had been introduced throughout the 1970s, and nothing had happened. Both felt the chances for getting concrete results, that is compensation, were better through a lawsuit.

After three months, Tome was getting nervous and worried. Udall had agreed to help but had not contacted him during the interim. Finally, in December 1978, Tome received a letter from Udall that told him an "action plan" was ready and that he was "anxious to get off to a fast start." The letter continued: "I can be at Red Rock on Tuesday, January 9, if such a date is convenient. My thought is that the best way to begin would be with a general meeting at your Chapter House with both the widows and the miners who are ill, followed by some individual interviews."[12] Udall and Mahoney made the trip and gathered files on about 200 miners and widows who would later become part of his lawsuits and eventually become his clients when the battle for compensation was won.

Suddenly in 1979 things began to happen. People from across the country were slowly becoming aware of the Navajos' problem. The environmental movement that Udall had quietly helped foster in the late 1960s spawned a new generation of social and environmental activists in the 1970s.

National awareness of the problems surrounding the nuclear weapons and power industry was heightened by the scandal that surrounded the death of Karen Silkwood, who had worked in a plutonium production plant owned by Kerr-McGee in Cimarron, Oklahoma. She had been contaminated with radiation because of inadequate exposure controls and unsafe working conditions at the plant, and was killed in an automobile accident on her way to deliver documents about these conditions to union officials and a news reporter.

There were many questions surrounding Silkwood's death, leaving the impression that the nuclear industry would go to any lengths, even murder, to cover itself. The net effect was to severely tarnish the once-lustrous image of the industry.

The industry's lack of concern for the people who fed its mines and mills was demonstrated by an organization called the Sunbelt Alliance, which was formed in the late 1970s in Oklahoma. Through its antinuclear activism, which

KAREN SILKWOOD POSES WITH HER CHILDREN, FROM LEFT: BEVERLY, DAWN, AND MICHAEL. SILKWOOD LATER DIED AFTER FILING A LAWSUIT AGAINST KERR-MCGEE CORPORATION.
PHOTO: AP/WIDE WORLD PHOTOS

included protests at the Black Fox Nuclear Power Plant in Inola, Oklahoma, and involvement with Native Americans, the group had become aware of the lung cancer deaths of about twenty-five uranium miners in the Red Valley area. The alliance organized a statewide food drive in Oklahoma for the fatherless families and delivered the food shortly before December 1978. In a gesture to dramatize the apparent disregard Kerr-McGee had for its workers, the food drive was launched with a letter hand-delivered to Dean McGee, a member of the Kerr-McGee empire, at his exclusive Oklahoma home on Thanksgiving day. The messenger got only as far as the kitchen, where a cook took the letter and said McGee did not want to be disturbed.[13]

About the same time that Udall visited the Red Valley community, so did freelance journalist Chris Shuey of the Phoenix area. Shuey said he was appalled at the lack of local media interest in the high death rate and health problems among the Navajo miners. He spent a day at the Red Valley Chapter House interviewing the widows and miners who were suffering from lung cancer and other diseases tied to their mining experiences. They told Shuey how the mining companies had never told them of the dangers of uranium mining and how, even after ventilation regulations were imposed, little effort was made to comply with them. Well into the 1970s the miners were forced to return to the mines within thirty minutes after blasting. Mine inspectors were taken to the cleanest parts of the mines to test for radon so that mine officials could continue their operations uninhibited by safety regulations.

Willie Begay, Sr., told Shuey how he was forced to push forty to forty-five one-ton ore cars a day up and out the long mine tunnel or be fired. In most other mines, donkeys or motorized cars performed this task. When a couple of Navajo uranium miners, Joe Tom Tapaha and Dan T. Benally, began to meet with other miners to discuss why they had splitting headaches, were nauseated, and frequently coughed up blood, they were fired by Kerr-McGee. Mrs. Mae John, a mining widow, told Shuey, "They would tell the Indians not to talk about the safety and go back into the mines or get fired. They treated them like animals."[14] Shuey had a difficult time placing the article for publication, but months of persistent effort finally paid off and his article appeared in the *Scottsdale Progress Saturday Magazine* on June 2, 1979.

While Shuey worked, the nation was shocked when malfunctioning equipment and operator errors at the Three Mile Island nuclear power plant in Pennsylvania released radiation into the air and the waters of the Susquehanna. The dangers of nuclear energy and the nuclear industry became dramatically apparent. Suddenly large populations were threatened by unknown quantities of radiation, and the nation itself felt exposed to dangers over which it had no control. The nation had a chance to feel what the Navajo miners were feeling when they learned that the mining companies and the government had knowingly sent them into very dangerous mines. In Red Valley, the Navajos were left to watch their friends and neighbors die as the rest of the nation panicked over a cloud of radioactivity drifting high over Pennsylvania.

In April 1979, Udall's case on behalf of the downwinders was gaining momentum. On April 19, Sen. Edward Kennedy conducted hearings on the downwinders in Salt Lake City for his Health and Scientific Research Committee. Udall had already filed claims on behalf of the downwinders against the Department of Energy (DOE), the successor to the Energy Research and Development Agency, which in turn had replaced the Atomic Energy Commission. Udall was claiming that the AEC knew of the dangers the tests posed to the general public and therefore was responsible for damages and compensation to people who fell victim to cancer caused by the fallout. He was correct in assuming that the Department of Energy would reject the claims, thus clearing the way for him to take the case to the legal system and attempt to get a judicial judgment against the government. Udall would take the same approach on behalf of the uranium miners.[15]

The plight of the Navajo miners began to receive more attention. In May

1979, *New York Times* reporter Molly Ivins made a trip to Red Valley. Much as Shuey and others had done, she told the story of the Navajo miners and how they were dying. The uranium miners' study was now being conducted by Dr. Joseph Wagoner, a Public Health Service researcher, and he gave her the death watch update. By 1974, 144 of the 3,415 Colorado Plateau miners that had been followed had died of lung cancer. Only about 30 such cancer deaths would have occurred in similar groups without such high radiation exposures, Wagoner told her. He estimated that the number of deaths would soon total 200, "making 160 people that needlessly died due to lung cancer because we did not accept the published data that was already there for our use."

Because Kerr-McGee had been one of the major mining companies in the Red Valley area, Ivins contacted the company for a comment on the Navajo deaths. She wrote that "A spokesman for Kerr-McGee said, 'We are not aware of the sources of the allegation that Navajo Indians who worked in those mines have died of lung cancer resulting from radiation exposure. We doubt that any evidence to support this allegation exists anywhere. Several efforts have been made to locate early uranium miners and trace their life histories in reference to the effects of radiation exposure. But none of these have been completely successful since miners change jobs frequently.'"[16]

What the Kerr-McGee spokesman was saying was that a large and powerful mining company was ignorant of the considerable statistical information that had been collected in the United States, beginning in 1950 with the Uranium Miners Study, and which had been presented over several months of public congressional hearings in 1967; the further scientific and medical studies which had been done in the intervening twelve years; as well as the information available prior to the 1940s from European uranium mining experiences.

Wagoner could only shake his head and restate what the scientists already knew: "From the point of view of scientific validity, there is no question about the cause of the deaths," Dr. Wagoner said. Another doctor, LaVerne Husen, who was in charge of the Indian hospital in Shiprock, witnessed on a regular basis the devastation that the mining was causing. "It got so I didn't need to wait for the tests," he said. "They would come in—the wife and the six kids were always there too—and they would say they had been spitting up blood. I would ask and they would answer that they had been in the mines. Then I didn't need to wait to make the diagnosis, really. I already knew. Those were the cancer cases."[17]

One readily available and public study that Kerr-McGee officials should have known about was done in 1970 by two doctors, Arthur R. Tamplin of the Bio-Medical Research Division of the Lawrence Radiation (Livermore) Laboratory and John W. Gofman of the Division of Medical Physics at the University of California in Berkeley. Titled *The Colorado Plateau:*

Joachimsthal Revisited, the study had been presented to the Joint Committee on Atomic Energy nine years earlier, on January 28, 1970, as part of the committee's continuing hearings on radiation and worker safety. Tamplin and Gofman, who went on to make their careers in studying the biological effects of low-level radiation, had reanalyzed the available statistical information on radiation exposures. The reason was that the old Federal Radiation Council, now free of critics like Willard Wirtz and now operating under the Republican administration of Richard Nixon, was considering a revision of the radiation exposure standard.

The opening lines of the 1970 study show that the questions that had once surrounded the correlation between radiation exposure in the mines and lung cancer had long since been answered. "There is no longer any significant debate concerning whether or not the inhalation of radon daughters in uranium and hardrock mining is associated with an excessively high incidence of lung cancer in the miners," Tamplin and Gofman wrote.[18]

Instead of repeating decades of previous work, they questioned the widely held assumption that low-level exposures did not carry the same risk that high levels did. The council had been operating under the assumption that steady work in a mine with an exposure of 1 working level, or 300 picocuries, the industry standard, carried no apparent risk of causing lung cancer. Gofman and Tamplin said the same information that always had been available showed that the risk was always present, even at low levels. Referring to the council's contention that total exposure under 1,000 WLM was safe, the two wrote:

> There would appear to exist no rational way to explain these statements. The same data were available *then* to the Federal Radiation Council as are available to us now for the analyses completed above. Our analysis showed clearly that radiation-induction appears to have an essentially constant doubling dose... down to the lowest dosage categories. The *deviation*, if significant, in the low dose categories was in the direction of a *higher* risk of cancer induction; *not* lower.... So there was clear-cut evidence available for radiation induction way, way below the 1,000 WLM spoken of by the FRC.... That we, in the United States, have repeated much of the tragedy of Joachimsthal and Schneeburg in the period from 1940 through 1967 is indeed regrettable. But human errors *do* occur, and it is best to forgive those that are past. However, if we go forward with a new generation of miners and repeat the Joachimsthal experience *again*, this will be absolutely inexcusable.
>
> It is our contention that the revised FRC Guidelines for uranium mining exposure to radon daughters (12 WLM per year) has [*sic*] a dangerously high probability of needless repetition of the Joachimsthal tragedy. (Even the *more* recent 3.6 WLM per year of the Labor Department is high.)[19]

The reams of studies and growing cadre of scientists and health officials who were preaching safety to the nuclear industry on a regular basis were accessible to anyone who wanted the information.

Meanwhile, as miners continued to die, their struggles for compensation were becoming more and more difficult, not easier. The town of Marysvale,

Utah, lay in the heart of Utah's rich uranium mining belt. Indian and non-Indian workers were employed in the numerous small and large mines that riddled the area. Today the town's small cemetery is the final resting place for so many former uranium miners that it is known locally as "uranium hill."

Hundreds of workmen's compensation claims were filed by the widows and descendants of dead uranium miners, many from the Marysvale area, but the Utah State Industrial Commission refused to recognize radiation exposure as a hazard of uranium mining and would not classify it as an occupational disease. Many widows objected to this because their husbands and the companies for which they worked paid the state regular premiums for the insurance; nevertheless, the state refused to make awards. The neighboring states of Colorado and New Mexico had already recognized the hazard and paid such claims. Utah was also sidestepping some claims by saying they had not been filed within the required one-year deadline after the worker's death. Utah's uranium widows appealed to Sen. Orrin Hatch, who promised to reintroduce yet another federal bill that would award compensation to widows of uranium miners. As had happened nearly every year that decade, a Senate bill was introduced but failed to get out of committee.[20]

By early July 1979, Udall and Mahoney were ready to file their initial claims against the U.S. Department of Energy on behalf of the Navajo uranium miners. Tome had worked closely with the two lawyers and the investigators they had assembled, including several of Udall's children. On the eve of the filing, Udall was optimistic, writing:

> Harry, you should know that we have uncovered new evidence in recent weeks which is very damaging to the government. I now believe your people have a strong case against the government, and I want you to know that we intend to use all of the skills we have to prosecute these cases as vigorously and aggressively as possible.
>
> Much work remains to be done, but in my opinion we have all done our homework well, with the result that the claims are in a much stronger posture today than they were a few weeks ago.[21]

On July 16, 1979, an unexpected event revealed the fallacies about safety in the uranium mining and milling industry. An earthen dam containing a large settling pond near Grants, New Mexico, where United Nuclear operated a uranium processing mill, gave way. The settling pond was filled with acidic and radioactive liquid wastes from the uranium ore refining process. Ninety-four million gallons of toxic wastewater and 1,100 tons of tailings spilled into the Rio Puerco de Oeste. The Rio Puerco flows through Gallup, New Mexico, and continues through lands used by the Navajos to graze and water their flocks of sheep and herds of cattle. The acid and radioactive waste gushed along the river for miles, permanently contaminating the land it fed and ruining the water for decades, if not centuries. (There are still signs along the river that warn of radiation.) The local press termed the event an environmental disaster. The U.S. Geological Survey called it the largest single release of radioactive waste in U.S. history.

On Friday, July 20, 1979, Udall filed thirty-two claims for $500,000 each in damages against the Department of Energy in Washington, D.C. Sixteen of the claims were on behalf of widows of deceased Navajo miners who had worked in mines both on and off the reservation. Eleven claims were on behalf of living uranium miners, some of whom were suffering from pulmonary fibrosis and some of whom were dying of lung cancer. Three of the claims were from widows of miners who had worked in the Marysvale, Utah, mine owned by the Vanadium Corporation of America, which also operated some mines in the Red Valley area. Of the fifty-five known miners who worked in the Marysvale VCA mine, twenty were dead. Two separate claims were filed by the son of a deceased Marysvale miner.[22]

The fight for compensation for the Navajos and other uranium miners had officially begun. The filing drew immediate attention from the news media in the West and spurred efforts by congressmen from the four western states affected by the problem. One of these was Sen. Pete Domenici, who scheduled a congressional hearing on August 30, 1979, in Grants, New Mexico, to hear about the plight of the Navajo and white uranium miners. The hearing officially was held by the Special Committee on Aging, chaired by Domenici, and the topic was "Occupational Health Hazards of Older Workers in New Mexico." At the time, Domenici was drafting yet another bill, dubbed by some the "white lung" bill, that would provide special compensation to the victims and families of stricken uranium miners. It was patterned after the black lung compensation bill for coal miners that had been passed earlier.

The hearing was supposed to be more than an opportunity for Domenici to make political hay, but he did not miss this chance to be in the limelight. Several reporters were on hand. In the glare of camera lights, Domenici toured an underground uranium mine near Grants in the morning, then convened his one-man hearing in a ballroom of the Grants Holiday Inn. The room was jammed with more than 300 people, many there to testify, some just to watch, and some to report. There was an air of hopefulness. Finally, someone had come to listen, even if it was just one senator. Finally, the miners could tell their story in their own backyard to someone who might be able to make a difference. The miners had good reason to be hopeful because Domenici was acutely aware of the problem and readily admitted that the government was at fault for endangering the lives of the miners.

He opened the hearing by accurately summing up the situation: "The U.S. Government, the sole procurer of uranium for defense purposes during the period 1948 to 1962, failed in its responsibility to make sure that uranium mining

FORMER SECRETARY OF THE INTERIOR STEWART UDALL BECAME AN AGGRESSIVE ADVOCATE OF COMPENSATION FOR DEAD AND DYING NAVAJO URANIUM MINERS. HE IS PICTURED AT THE JUNE 5, 1993, CONGRESSIONAL HEARING IN SHIPROCK, NEW MEXICO.
PHOTO: MURRAE HAYNES

operations were carried out in a manner which insured the safety and health of the workers involved." He went on to admit that the AEC knew of the dangers in the mines and even solutions to the problem, but took no action, leaving the miners to become the "early victims of the nuclear age."[23]

The senator's words were more than a little gratifying to people like Harry Tome, whose ten-year struggle for compensation for his friends in Red Valley had been largely responsible for the current hearing. In the room waiting to testify and to listen were many of the widows, children, and cancer-stricken miners whom he had been counseling and for whom he had been working.

Testimony began with short statements and questioning of seven white miners. Then Agnes M. Ratliff, the widow of a man who had worked in one of the larger mines in the Grants area, suggested that mining companies pressured the local Grants clinic, which examined almost all the uranium workers in the region, to withhold the results of the deadly conditions under which they had worked. Ratliff's husband, Charles, died on August 13, 1975, at the age of forty-six. He had worked in two sections of the United Nuclear Homestake mine for seventeen years. Ratliff was a textbook case of lung cancer from uranium as found in the hundreds of cases in the European mines. He fit the age group, the mid-forties, and became ill after seventeen years in the uranium mines. Also, his death had come swiftly.

He first became ill, his widow told Domenici, on July 18, 1975, and immediately began to lose weight and his appetite. Six days later he was admitted to an Albuquerque hospital and underwent radiation treatment and tests. His condition did not improve so he was sent home. Little more than a week later, on August 10, he collapsed in a coma and was returned to the hospital. Three days later he was dead. An autopsy showed that he had lung cancer and that "the cancer had spread all through his body."

Under questioning by Domenici, Mrs. Ratliff said that her husband had had twice-yearly checkups that included X-rays at the Grants clinic, which was where all the uranium miners' records were kept. After her husband died, she sought his medical records, including his chest X-rays. The clinic doctors refused to release them, and Mrs. Ratliff was forced to get a court order to obtain them. The X-rays showed the dark spots on his lungs that are generally interpreted as silicosis or as precursors to lung cancer. "The doctor told my husband at that time [in Albuquerque] they were just sick to see his lungs, the condition they were in," Mrs. Ratliff said. Domenici asked if the checkups were regular. Mrs. Ratliff answered, "Yes; I believe he took two a year. On his last one, they had called him back in and took X-rays and then his foreman had told him there was nothing wrong with him, and that [was] in March of the same year."[24]

The story was repeated frequently. Speaking on behalf of his dead father, Dennis Heppler, who was also a miner, said that his father had died of cancer but that not long before his death he had been told by the mining company that nothing was wrong. "He received an annual physical about 4 months before he died and they found nothing wrong with him. He was feeling bad and we took him to the veterans hospital. He had cancer and he was dead in 6 weeks."[25]

The hearing quickly moved to a panel of attorneys, including Stewart Udall. Udall was characteristically frank. Lawsuits were being filed that day in Salt Lake City on behalf of the downwinders, he said. Despite that tragedy, Udall said that after a year of research on the uranium mining issue, he was even more depressed. "I have seen a lot of buckpassing in the Government in my day, and I must say this is the most outrageous example that I have ever seen because the different agencies of the Federal Government each shrugged their shoulders, each passed the buck on to someone else and it wasn't until about 2 or 3 months ago that I began to see daylight in my investigation."

Udall said that it was now clear that the Atomic Energy Commission knew the miners would die but failed to act. He noted, however, that the government had surrendered its previous immunity from prosecution in what was known as the Federal Tort Claims Act of 1946. This act allows private citizens to seek remedies in federal courts for injuries they feel have been incurred by the failure of the government to take reasonable, due care in its operations. (However, as we will see later, there are important exceptions.) Knowing that the government could well be held liable, federal officials should have acted to warn both miners and mining companies of the dangers of uranium mining, thereby protecting the government against future suits, Udall argued. "They [the agencies] denied they had responsibility, but the Federal Tort Claims Act was passed in 1946, and, in my view, they had a duty to warn the miners, to warn the mine operators, to warn the states and the state safety agencies, to warn the Navajo Tribe, to warn the Bureau of Indian Affairs, of the dangers that they knew about."[26]

Udall was sensitive to the possibility that legal actions in the courts and legislative actions in Congress might stall each other, because occasionally one branch of government will not act until the other does or refuses to. "I hope that you don't regard the efforts that I and others may be making through the court system…as counter productive. It may be that this is the way to get the job done." Unfortunately, it did not work out that way, and Congress ultimately failed to act until the issue had been fully litigated.[27]

Udall was critical of Domenici's proposed legislation. First, it was not clear that it would provide compensation to past victims, although it was clear that all future victims would be aided. Second, the amount of compensation was relatively small, ranging from about $150 per month to an individual survivor to about $350 for a family. Considering that miners had paid and would pay for their work with their lives, it seemed a paltry sum, and this irritated Udall. He pointed out that the government seemed more than ready to spend huge sums of money on projects, such as cleanups, but not on people.

About $100 million was being spent on cleaning up a Pacific island where bomb tests had taken place, even though decontamination was practically impossible. Just a year earlier, in 1978, the government had agreed to spend $180 million to cover up the uranium mill tailings piles that were scattered across the West in two dozen locations and which, in some cases, had been left exposed for nearly thirty years. "Why is it that many people get horrified if someone proposes that we spend $20 million, or $50 million or $100 million helping the families and survivors of these uranium miners? I think it's about time that we faced up to the fact that there would have been no nuclear arsenal if there had not been these human beings who were willing to make sacrifices for their country."[28]

Tome led the charge on behalf of the Navajo uranium miners. In his prepared testimony, he painted a bleak picture and revealed that the dangers of the uranium mines on the Navajo Reservation extended well beyond the lung cancers that were striking the miners. He said the average annual income for a Navajo miner's family was only $5,184. The families averaged four children, and when the husband became ill and was unable to work, the families slipped deeper into poverty. Families of the miners also developed serious health problems, he said. One wife of a miner inexplicably delivered four stillborn babies.

Tome told how a crew from CBS television came to do a story on the stricken Navajo uranium miners a few months before the hearing. The crew was guided by Harold Tso, director of the Navajo Environmental Protection Commission,* to some abandoned mine sites. "On their way, they stopped to

*The predecessor of the Navajo Environmental Protection Administration.

talk with a miner and his family. A member of Mr. Tso's office accidentally turned on a Geiger counter, and it let out a loud sound. Upon further investigation it was discovered that the house, which was made out of rock, was made out of highly radioactive material."[29] The house had a constant reading of about 100 rems of radioactivity. It was the equivalent of living in a uranium mine. Ultimately, the tribe would discover that twenty-two homes in the Red Valley area and dozens more across the reservation were made of either mine waste or mill tailings. Since the uranium-bearing sandstone fractured into square and rectangular shapes, it provided a readily available and usable source of building materials. The miners were unwittingly exposing their families to the same kind of radon and radiation bombardment that they were subjected to in the mines. Eventually, the Navajo tribe replaced seventeen radioactive homes in the Red Valley area with the aid of federal money, but the long-term health effects from the exposure the families suffered have never been studied.

One of the miners who had built a house of the radioactive rock, Carl Thomas, testified at the hearing. A Navajo from Red Valley, Thomas began working in the uranium mines in his neighborhood in 1947. He told Domenici that at the time, there was no ventilation in the mines. He quit working in the mines twelve years later in 1969 when he became ill. He was suffering from lung cancer at the time of the hearing and explained to Domenici that because of the poverty on the reservation, the rock was the only available building material he could afford: "I built this house and I didn't know that it contained high radiation, uranium. I still live in that house. I was told that it is very dangerous to live in this stone house. I don't have any money. I don't work. There is no way I could get money to build me a house that does not contain radiation from uranium."[30]

Speaking on behalf of her mother, Pearl Nakai, and her deceased father, John Smith Nakai, who had been a uranium miner for twenty-three years, Marie Harvey told a tragic tale. Her father had worked in uranium mines in Colorado, Arizona, and Utah, and besides breathing high levels of radon, regularly drank the cool water that dripped from the walls of the mines:

> At the time that he was working, we were never told about the conditions or were we cautioned about what was going to happen to him. They drank the water that seeped out of the walls of the mines. We lived about 50 feet away from the dumps of the ore. As kids, we were never told not to go here and there, or play here and there. Now my father's clothing, he took it off, he hung it in the house. We lived in a one-room house there. Nobody told us about the dangers of the uranium ore until it was 1974, and we found out he had cancer of the stomach and the liver. There was some in his lungs, too.... He had an operation and...it was only 6 weeks for him to live after he was operated on. We found out he did have cancer. He had worked with 30 men and most of them have died; some of the widows and children are sitting in the back today. They [would] testify also, if they could, about the conditions in the mines.[31]

Jessie Harrison, a uranium miner's widow, told the story of her husband's battle with cancer in her own simple English:

> I am a widow too. It is hard on us to talk about it. Sometimes I don't want to talk about that. It is new to me. It is hard on me to think about it. It is the same for me as she said. I am a widow. I have six kids, three boys and three girls. It is hard on you to help the kids without your husband. I know that he was very, very sick and he died. We didn't even know this thing about the mines. He was working in the mines 19 or 20 years. He was working over at Kerr-McGee at Cove, Arizona, about 16 years. Then we left Vanadium Corporation of America and he worked there 3 years at Cove, Mesa. We moved to Colorado, Gateway, Colorado. We were there about 5 years.

She explained that her husband came down with a severe cold in December 1968. Although he attempted to return to work in Colorado, he could not work and finally went to the Shiprock hospital in late January 1969. The X-rays showed that cancer was in his lungs and throat. The family was in shock, knowing well what that meant and angry at the mining companies that had given him regular checkups but had told Harrison nothing. On February 15, he had most of one lung removed in an Albuquerque hospital.

> He came back from the hospital. The doctor told us if he passed 8 months, he was going to be all right. If not, he is not going to be getting well. If he goes past 4 or 5 months, he is going to be all right. We just prayed and prayed about that. It was summer and August, almost the end of August he started in bleeding again from his mouth. We just go down, everybody go down. He tried to eat, but he just throw up. I just took care of him all the time. I didn't ever think about me. I don't eat good and I don't feel better. I don't eat good. I don't buy anything. I am just working.

Jessie Harrison's husband died in January 1971.[32]

Harris Charley, a Navajo miner from Beclabito, New Mexico, told Domenici about the conditions in the mines. He worked in uranium mines for twenty-eight years in Arizona, New Mexico, and Colorado. He was sick, had trouble breathing, had a lot of pain and trouble sleeping:

> In those days, miners were not taken good care of. No air. Old mines caving in. No safety was there. We were just kicked around, treated like dogs. Using the water to drink in the mines, inside the mines. Lots of hard labor, 8 to 10 hours, day and night. The company only wants more uranium to be shipped out and more money. Doesn't really care for human beings. The mine inspector was there around 1955 and we started having air to reach into mines. I think this is very dangerous for a human being. We don't like our grown children to get into this kind of disease. I am not under any kind of medication. The doctor has told me just to take care of yourself, that is the only way, then, you are going to live long. I kept on praying to the Lord.[33]*

The hearings were by most standards a success. Coverage in the news

* Harris Charley also provided testimony at a later hearing in April 1982. However, he was too sick to deliver it, and his son Perry Charley read it for him. He went into full respiratory arrest on October 15, 1982, from the effects of fibrosis of the lung and was in a coma until his death on December 19, 1986. One of the original ninety-five miners in Udall's lawsuit in 1979, he had received a cumulative exposure of 1,192 WLM.

media was extensive, locally and nationally. Clearly, the story was out. Udall had already filed claims against the Department of Energy, anticipating that they would not respond within the six-month waiting period, thus clearing the way for his lawsuits against the government and the mining companies. If the government did not respond to the damage claims, then Udall could argue that all possible administrative remedies had been exhausted. In early September 1979, Udall filed 185 claims against the government on behalf of 22 uranium miners, 25 widows, and 138 of their surviving children. The claims totaled $44 million.[34]

Meanwhile, more congressional legislation was being drafted while still more bills were being introduced as a way to address the problem. Domenici's bill was finally introduced in December but went nowhere because the holiday recess was at hand. Harry Tome once again appealed to the Navajo Tribal Council, urging it to pass yet another resolution calling for the U.S. Congress to act on a bill to provide compensation to the uranium miners and their survivors.[35]

On December 15, 1979, Udall, who was anxious to get his suits moving, filed a lawsuit in federal district court in Phoenix on behalf of ninety-five Navajo miners, widows, and their descendants against seven major mining companies on the Navajo Reservation, including the Kerr-McGee Corporation; Kerr-McGee Oil Industries; Vanadium Corporation of America; Foote Mineral Company; AMEX; Climax Uranium Company; and Climax Molybdenum Company. In the suit, Udall claimed that the mining companies knew of the dangers in the mines but did nothing to warn or protect the miners. He called it a "breach of duty" on the part of the defendants and asked for millions in damages. He asked that each of the people listed in the complaint receive $500,000, that each widow receive an additional $500,000, each living spouse (widower) $150,000, and each child $150,000.

With the filing of the suit, a long legal battle began. The total award requested was $30 million plus unspecified damages. Udall and Mahoney were upbeat about the prospects of the suit because of the success of the Silkwood case against Kerr-McGee, which obtained $10.5 million for one family. In an interview with a *Washington Post* reporter, Udall said, "It is a theory of law that says that if an activity is abnormally dangerous, all you have to show is that it is abnormally dangerous and that the defendants knew that. You don't have to prove actual negligence."[36] Udall also explained that in addition the companies did not comply with workmen's compensation laws because the dangers were not posted in the Navajos' own language. Few if any understood English, and even fewer could read and write it. In the same article, W. E. Heimann, the general counsel and vice president of Kerr-McGee, would not comment on the suit, saying he had not seen it.

Two months later in early 1980, the uranium miners' compensation bills that had been introduced in Congress showed little signs of life. A delegation of

Navajo uranium miners and widows, including many who had testified at Domenici's August hearing, were brought to Washington, D.C., by a group called the Citizens' Hearings for Radiation Victims, a coalition of environmental activist groups concerned about radiation-linked illness. They held a press conference in which the Navajo delegation delivered its public appeal for compensation. The event received widespread coverage, and the story was distributed nationally by the Associated Press.

Meanwhile, medical evidence that radon daughters caused lung cancer continued to mount. Dr. Joseph Wagoner told a group of doctors in March at the University of New Mexico Medical School that smoking did not necessarily have to be linked to radon before cancer was produced. Wagoner said recent studies of Swedish uranium miners who were nonsmokers showed that these miners generally came down with lung cancer as long as twenty-five to forty years after their exposure. These studies were among several that went a long way to destroy the widely held assumption that only uranium miners who smoked were susceptible to lung cancer. They also reinforced other work that had involved Navajo uranium miners. The uranium mines prior to 1962 were poorly ventilated, Wagoner said, and the Navajo uranium miners generally did not smoke. However, in a study of about 700 uranium miners, 11 died of lung cancer when only 2 or 3 would have normally died of it. Navajos, as had the German and Czechoslovakian miners, showed signs of cancer from seventeen to nineteen years after exposure. However, for miners who were heavy smokers, cancer appeared as early as twelve years after exposure, Wagoner told the group.[37]

Back in Red Valley, Harold Tso struggled to get new housing for the miners in the area who had been living in radioactive houses. Tso complained that the Navajo tribe had no money to replace the contaminated homes. He wanted to relocate the afflicted families, but, he said, the families themselves did not want to move because they had developed a deep attachment to the place where they were born. While the homes in Red Valley had been built with radioactive waste rock from the mines, the homes in Cane Valley had been built with concrete that used the gray, rough sand that was left over from the conversion process—the mill tailings. Even in Grand Junction, Colorado, where the AEC was headquartered and uranium mills had been located, large amounts of mill tailings had been used to build houses, schools, and government buildings.

Tso was already at work trying to get money from the tribe's fees collected from coal mining under the Surface Mining Control and Reclamation Act (SMCRA) to begin cleaning up hundreds of abandoned uranium mines on the Navajo Reservation so the radiation problem would spread no further.* This

---

*When the Surface Mining Control and Reclamation Act of 1977 was passed, it focused on the problems of the eastern states, which were those of coal mines. The regulations did not address the problems of uranium mines, and so, before any money can be used to clean up uranium mines, Navajos must show that the uranium mines present an extreme health hazard.

was the first such attempt to clean up the abandoned uranium mines which riddle the West, and the Navajo Abandoned Mine Lands Reclamation Department was established in 1988 as a result of this effort. The U.S. government still does not have a similar program for uranium mines.[38]

While Tome continued to write to the U.S. Congress, Tso pressed for housing and Udall fought in the courts. For Udall, his battle had become something of a labor of love. He had enlisted the help of his family to conduct the necessary research. His daughter Lori and sons Denis and Tom (the latter is currently the attorney general of New Mexico) had been doing a lot of the research on a voluntary basis. Udall knew he was up against some stiff competition and high-quality and high-priced lawyers working for Kerr-McGee. Having been stung by the award made in the Silkwood case, Kerr-McGee was taking no chances this time.

Udall appealed to the tribe for some financial help in a letter to former Navajo tribal chairman Peter MacDonald:

> Peter, in the coming months we are going to be at a great disadvantage in our court battles with the defendants if we cannot find a way to raise enough money to compete on even terms with the government and Kerr-McGee and the other powerful mining companies. There will be heavy travel, deposition costs, etc. as the defendants have already indicated they intend to take the deposition of all the Navajo widows and living victims. I am trying to get a benefit concert and some other fund-raising going, but I am also wondering whether there is any possibility that the Tribe might be able to appropriate say $10,000 to pay the travel and deposition expenses of the Navajo plaintiffs?[39]

Unfortunately, Udall was unable to generate a lot of support for his request because other Navajo Tribal Council members said the legal fees were supporting only a small segment of the tribe. However, in February 1980, the council did pass a resolution urging Congress to pass Senate Bill 1827, titled the Uranium Miner's Compensation Act of 1979.

Meanwhile, the search for radioactive homes in the Red Valley area continued. A series of meetings were held to inform and educate the affected families. These meetings were organized by Sarah Mae Benally and Perry Charley, who had been active in testifying before Domenici's committee and Congress on the problems that the Navajo uranium miners faced.

By the fall of 1980, all parties involved in the compensation issue were getting nervous. Legislation in Congress was going nowhere, and the lawsuits, as far as the Navajo miners were concerned, were essentially dead, too. Udall and Mahoney heard that their Navajo clients were being told to forget the suits. Their only hope was in Congress. Mahoney quickly fired off a letter to Harry Tome, hoping to stave off any future problems: "In our opinion, it would be a great mistake for any of our clients to take this drastic step. As we mentioned to you in Red Rock, both approaches should be followed simultaneously." Mahoney argued that the courts could be the source of larger settlements in the cases than would be provided by Congress. A congressional solution should be a fallback position, he wrote: "Thus, although congressional relief might be years away and would probably not provide the sizable awards we think can be obtained in lawsuits, it should provide a very sound guarantee of insurance for the future."[40] It was prescient advice because Congress would take another decade to act.

Udall said in a September 1993 interview that he had given the issue serious thought, but the more he learned of the tragedy, the more he was convinced that a lawsuit was the proper approach. "When we went to the Red Valley Chapter and did the interviews," Udall said, "you saw the tragedy and the scope of this thing." Having begun his legal career as a personal injury lawyer, Udall said, "I began to look at this as a personal injury lawyer. It was a challenge. I didn't correctly assess how much time it would take." On a personal level, Udall found the case also a reason to finally leave Washington, D.C., after twenty-five years. "It was a challenge and an opportunity to go back home."

In April 1981, Udall received some bad news. Despite strong lobbying efforts by Tome and other representatives from the Red Valley area and other affected communities, the Navajo tribe's comptroller disapproved a request for $15,000 to defray the costs of both Udall's suits and continued demands on the Navajos to travel to Washington, D.C., to testify on the compensation bills. In an appeal of the decision, the Red Valley Chapter stated that some 650 Navajos were stricken by disease or other problems stemming from the uranium mining and that "Funds are badly needed to pay for expenses of Navajos who will testify in court and in Senate hearings on uranium issues."[41]

Disappointed but not deterred, Udall continued the fight, financed by his writing and lecturing. In the fall of 1980, almost six months earlier, Udall's suits against Kerr-McGee, VCA, and Climax had been thrown out of court by the federal district court judge. This was no surprise to Udall, who immediately filed with the federal appeals court in San Francisco and by mid-May had submitted the necessary papers.

The mining companies successfully argued that the Navajo miners were covered by Arizona workmen's compensation laws and because of that could not file separate lawsuits for additional compensation. Udall argued that because the mines were on the Navajo Reservation, which was under the jurisdiction of the federal not the state government, the miners were not covered by Arizona laws. The judge agreed with the mining companies. Udall recalled later that the mining company lawyers had been able to get some Navajo workers who were part of the suit to admit that they may have seen a workmen's compensation notice posted somewhere on the mining company property. Udall said that the element of doubt introduced into the miners' minds, probably because the miners often worked other jobs and had probably seen such a poster someplace, was enough to prompt dismissal of the suit.

However, Udall did not give up. His suit filed against the U.S. government was awaiting a decision in mid-June 1981. With the help of his widespread contacts and several environmental groups, Udall obtained $15,000 to cover expenses from an organization called The Youth Project, a nonprofit group that adopted the Navajo miners and widows as one of their special projects. Also, the famous Navajo artist R. C. Gorman gave Udall the rights to a lithograph. In addition, Udall was successful in arranging a benefit concert at the Santa Fe Indian School in September 1981. The concert featured the famous folk singer Pete Seeger and environmental writer Edward Abbey, and raised about $7,000. This money was used to produce an extensive run of the Gorman lithographs, and the sale of these raised about $45,000 for the case.[42]

By mid-April 1982, yet another bill had been introduced into Congress, Senate Bill 1483, which was titled the Radiation Exposure Compensation Act of 1981. The Senate Committee on Labor and Human Resources held a field hearing at Salt Lake City, Utah, on the bill, which would award compensation to both downwinders and the uranium miners on the Colorado Plateau, including the Navajos. Tome testified, as he had done many times in the past, and reiterated the health and financial problems that the uranium mining had caused the Navajos.[43]

In the meantime, Udall had argued his case against the mining companies in the San Francisco federal appeals court and awaited the court's decision. In a newsletter to his clients, Udall, perhaps sensing the darker possibilities of the case, said, "If we lose, it is our intention to appeal your case to the United States Supreme Court in Washington, D.C."[44] In his other case against the government, Udall had secured an agreement from Federal Judge William P. Copple to select eleven representative cases for trial instead of all of the ninety-five claims that had been filed. Udall was still optimistic.

The case against the government was scheduled to go to trial in August 1983. More than a year earlier, Udall had been joined in the effort by bilingual Navajo attorney Albert Hale, who provided valuable language skills by reducing the amount of time needed to interview clients and discuss their testimony. Hale recalled how difficult it was to discuss the issue with many of his fellow Navajos. "It was a real problem because so many of them not only couldn't speak English, but just had no comprehension of what was going on, what had *happened*," he recalled. The Navajos have no word for radiation, and the lawyers used the concept of steam in their attempts to explain the dangers of an odorless, colorless, tasteless substance, Hale said.[45] The bilingual capability that Hale provided was invaluable, Udall said in 1993. "In preparing these Navajo witnesses, he was very important."

At the trial, and as had been agreed, Udall put his eleven plaintiffs and a host of expert witnesses on the stand, including Duncan Holaday and Victor Archer, whose testimony would show that the deaths of the miners from lung cancer could have been prevented because the government knew of the dangers, yet had taken no action. The solution to the problem was simply ventilation, as Duncan Holaday had demonstrated in his interim report in the early 1950s. The government and the mining companies knew of both the problem and the solution but had failed to act, Udall said. "This is a case of those with the power to do something who looked the other way. That is the tragedy," he said in his opening statement.[46]

The government's counterargument was that it was exempt from blame in the case because it had exercised its "discretionary function" in the situation. This discretion, the lawyers contended, allowed the government to make certain decisions to carry out programs despite some possible health consequences.* From Copple's early remark and later through the trial, Udall sensed that he was going to lose. "It was clear he [Copple] was in the government's pocket," Udall said in a 1993 interview. "Every argument the government made, he bought. It was devastating."

In July 1984, Judge Copple ruled in favor of the government. He admitted, however, that the case clearly called out for compensation to be provided to the victims, but that in this instance, the government could not legally be held liable.

Udall was deeply saddened by Copple's decision because just two months earlier, in *Allen vs. the United States*, which was called the "downwinders' case," Federal District Court Judge Bruce Jenkins in Salt Lake City issued a ruling that held the government negligent for not warning the downwinders of the possible dangers to their health that massive amounts of fallout presented.

Both cases were appealed, of course, by opposing parties for their own obvious reasons. Udall appealed his case to the Ninth Circuit Court of Appeals in San Francisco and later went there with his son and co-counsel, Tom, and Albert Hale to argue the case. Suddenly, it seemed there was hope. Udall recalled that during arguments, the Department of Justice attorneys defending the government hammered away at the position that the Atomic Energy Commission was concerned but was waiting for the results of the uranium miners' health study to find out the facts about the dangers of radiation in the

---

*There are twelve exceptions to the Federal Tort Claims Act. The one invoked in this case, the so-called "discretionary function," is written in language so convoluted that it is still the subject of argument by scholars and jurists. In the miners' case, the court chose to interpret it to benefit the government.

mines. Somewhat irritated, one of the judges warned the justice department to drop that point. "Counsel, I would not dwell on that. You were using them as guinea pigs," Udall quoted in a later interview. Upon hearing those words, Hale and Tom Udall smiled at each other, sensing that the case had been won. However, in August 1985, Copple's ruling was upheld by the federal appeals court. In issuing its decision, the court said, "We agree with the district court that this is the type of case that cries out for redress, but the courts are not able to give it; Congress is the appropriate source in this instance."[47]

In the meantime, the Department of Justice appealed the downwinders' case to the Tenth Circuit Court of Appeals in Denver. In April 1987, the appeals court reversed the Jenkins ruling and sided with the government, arguing that it had exercised its discretionary function in this instance. Both the downwinders' and the miners' cases had in effect been won by the Department of Justice. Udall's only hope was the U.S. Supreme Court. Both cases were appealed. A year later in 1988, the U.S. Supreme Court, after reviewing the cases, refused to hear either and in effect upheld the lower courts' decisions that the government was not liable. It was the end of the road, and Udall was devastated. It was a benchmark decision in many ways, Udall recalled. It in effect set a new precedent for the government's immunity, despite what the Federal Tort Claims Act had stated in 1946. "That was just a horrible break we got," he commented.

The only thing left to do was to follow the court of appeal's advice and to once again go to Congress. Udall was not at all excited about returning to Congress. "I was so drained out then. I didn't think this thing [compensation] had a prayer," he said later. After all, there had been attempts for nearly twenty years to get Congress to do something for both groups. However, now the situation seemed to be a little different. Congress had heard and reheard the issue, and there were few that did not know about it by now, given the amount of publicity that the issue had received. Congress, in general, was more sympathetic. Also, Udall still had a lot of friends in Congress, particularly Rep. Wayne Owens of Utah, who as a lawyer in private practice had assisted on some of the Marysvale uranium miners' cases. Owens agreed to help push the bill. Later, he explained why Congress had been reluctant to act: "It is very difficult to get legislation through if you are fighting a judicial issue. Congress liked to think of itself as the body of last resort."[48]

Owens went to Sen. Orrin Hatch, who had been involved in previous efforts to get similar compensation bills through Congress. Hatch agreed to help on the Senate side if a bill could get through the House. Another factor in the Navajos' favor was that Congress had just passed a bill that provided compensation to the Japanese Americans who were placed in American concentration camps across the West during World War II. One of the key committees that had held a hearing on this bill was a House judiciary subcommittee chaired by Rep. Barney Frank of Massachusetts. He conducted a hearing on Owens's new bill in September 1988, but Udall did not attend. He was exhausted, emotionally and financially. "I didn't go. I didn't have a plane ticket," Udall explained. However, he was deeply involved. An early version of the bill removed the government's immunity in the radiation exposure cases and allowed the Navajos and downwinders to obtain remedies in the court system. Udall objected violently to this, saying that they had *already* been through the courts and did not want to go back. They were back in Congress because the courts had said no.

The next year Owens's bill was reintroduced as H.R. 2372 (S. 841 in the Senate) and again went before Frank's subcommittee. Udall was there to testify and was optimistic when the committee's ranking Republican talked to him and indicated his support for the bill, which meant it had bipartisan backing. Opposition to the bill came from the justice department, which said, as it had during the extended court battles, that the government should not be forced to pay for what had happened. The bill, however, looked as if it actually might move.

On March 13, 1990, New Mexico Sen. Jeff Bingaman conducted a hearing in the high school gymnasium in Shiprock to generate support for the bill. The hearing was titled *Impacts of Past Uranium Mining Practices* and was conducted by the Subcommittee on Mineral Resources Development and Production as part of the Senate's Energy and Natural Resources Committee.

The hearing was brief and to the point, more of a formal gesture than a true investigatory event. The facts and history of the Navajo uranium miners were well known. A gross wrong had to be righted, and this hearing would place a long-awaited stamp of approval on a bill that might finally be passed. Bingaman was joined by Sen. Pete Domenici, who talked about his longtime involvement in the issue and outlined the provisions of the proposed act. A $100 million trust fund was to be established from which awards of $100,000 would be made to miners who had worked in uranium mines beginning on January 1, 1947, and ending on December 31, 1971, who had been exposed to 200 or more working level months of radiation, and who had contracted lung cancer or some other serious respiratory disease. If the miners had died, the award would go to their widows or children. In addition, people who had lived downwind from atmospheric testing of nuclear weapons as well as the "atomic veterans," the soldiers and sailors who had been purposely exposed to radiation during atmospheric testing, would be eligible for $50,000.

The hearing began with a statement by Leonard Haskie, who was the interim chairman of the Navajo Tribal Council from February 1989 until Peterson Zah was elected president in November 1990. Haskie had replaced former Tribal Chairman Peter MacDonald, who had been charged in connection with tribal land purchase scandals. (In the eleven years that Udall worked on the court cases, MacDonald was unable to provide any assistance from the tribe.) To some extent, Haskie had been responsible for the hearing and the progress of the legislation. As one of his first moves, he reestablished the Navajo office in Washington, D.C., which immediately began to work hard on drafting and lobbying for the Navajo miners' compensation bill.

Haskie was doing a lot more than simply saying that the miners needed help. He told the two sympathetic senators that Congress was now the only source of hope since the courts had said they would not help. He reminded those at the hearing that the average annual per capita income for the Navajos was only $2,400 and that unemployment was at 32 percent. Many Navajos lived in primitive conditions, without roads, electricity, running water, or telephones, and with few hospitals. "I appeal to your compassion and sense of justice and ask you to make a compassionate gesture to those whose health was considered secondary to the national security interests of this country," Haskie said.[49]

Dr. Richard Hornung, chief of statistics for the National Institute for Occupational Safety and Health, provided an update on the medical studies. Although Willard Wirtz had ordered a standard of a 0.3 working level per month or 3.6 working level months per year in 1967, the standard was not accepted by enforcement agencies. In 1970, the Bureau of Mines instead established 12 working level months per year as the federal exposure limit for underground mines. In 1971, the Environmental Protection Agency issued radiation protection limits of 4 working level months per year. The 4 WLM standard was adopted by the Bureau of Mines in 1976. In 1987, the National Institute for Occupational Safety and Health was recommending that the standard of exposure be dropped to 1 working level month per year, which is less than a third of what Secretary of Labor Wirtz had ordered but had been unable to enforce.

By now, the facts on radiation exposure were beyond dispute. Hornung reported that the study of uranium miners showed that the miners were five times as likely to develop lung cancer as the general population. Of the eventual total of 4,146 uranium miners who had been studied, more than 350 lung cancer deaths had been reported. At that rate, one of every twelve miners, at the exposure standard of 12 WLM per year, could be expected to die of lung cancer.

Hornung warned that the existing standard of 4 WLM per year still did not guarantee safety. Continued scientific study on the subject of exposure confirmed what Drs. Gofman and Tamplin had proved in the 1970s: there is no safe level of exposure to radiation. "In other words, this means that there is no totally—or no totally safe level of radon exposure is assumed, and that each additional WLM of exposure above what is naturally occurring background will produce a small increment in lung cancer risk. Therefore, one must speak in terms of acceptable risk, rather than a totally safe level of exposure." Hornung said that even at the low rate of 4 WLM per year, an accumulated thirty-year dose of 120 working levels would double a uranium miner's chances of lung cancer.[50]

The impacts of uranium mining and milling went far beyond physical health. Susan Dawson, a doctor of social work at Utah State University, said that the psychological strain of living around known health hazards like uranium was devastating to Navajos, who held a view of the earth as sacred and nurturing, not deadly. The deaths from radiation exposure were clearly destroying family and social life:

> Psychologically, the overall impact of the uranium mining and resulting health and environmental problems upon the Navajo have widespread implications. They view themselves and their world as an interconnected whole. This applies to religion, concepts of health, and how they perceive themselves in relation to the world. Consequently, a disruption in one part of their lives, be it their health or the land, creates an imbalance or disharmony in the overall system. The disruption, then, causes stress, not only because of the immediate problem to the individual, but because it threatens the disruption of the Navajo fabric of life and wholeness.[51]

Phillip Harrison testified on behalf of the families of the miners who had died. The son of Phillip and Jessie Harrison, a widow who had testified at similar congressional hearings more than ten years earlier, he spoke on behalf of the Uranium Radiation Victims Committee, a citizens' group that had been formed to help victims and their families. The decades of pain and suffering that the uranium mining had caused and the resulting resentment showed in Harrison's impassioned words:

> Through the years, Navajo miners were guinea pigs for the Public Health Service and the Bureau of Mines. Mr. Chairman, the companies' and the Atomic Energy Commission's sheer negligence was uncalled for, and the recognition for the Native American miners is long overdue.... After many years of frustration, hardships, pain and suffering, we are entitled to be recognized and compensated for our losses. The compensation cannot replace our loved ones but will relieve and remedy our burdens. But our work cannot cease. We have yet to address our land, water, livestock and food cycles that are contaminated.
>
> We also have our children that are genetically affected from the impact of uranium mining and spouses that are suffering from the same symptoms as their husbands. We also share our sympathy with the Utah downwinders, atomic veterans, mill workers and all the radiation victims worldwide.[52]

There were new legal reasons for the bill to be passed. Earl Mettler, an attorney in private practice in Shiprock, described how the state workmen's compensation laws had become a bureaucratic maze for Navajos who attempted to get payments for deaths and disabilities. In Colorado, for example, disability payments were often as little as $21 over and above what Social Security was providing. The miners were in a Catch-22 situation. They could not receive both Social Security and compensation, and often a successful lawsuit would produce no more money than the miner was getting from Social Security. In addition, compensation was based on wages paid at the time the disability was incurred, and Social Security was based on payments in effect at the time of filing. Since some of the cases took years to resolve, this meant that the payments finally awarded were usually inadequate for current needs. Some states did not allow claims for any diseases that occurred more than ten years after employment. Since the latency period for lung cancer from radiation was often twenty years, this prohibited the claims from being paid. Mettler noted that this effectively destroyed the justice department's contention that compensation for the miners was already available.[53]

Finally on June 5, 1990, after a lot of lobbying, which included a letter from Udall to key congressional committee members that tore apart the Department of Justice's opposition to the bill, the House passed it. Then, two months later on August 1, an amended bill was passed by the Senate. However, Senate passage of an amended bill was both good and bad news. The fate of the bill hung in the balance. Normally if the two houses of Congress pass different versions of a bill, a conference committee is appointed to work out a compromise, which then must be approved by both houses. At this point, there are still many ways for a bill to die. The conference committee could fail to agree on a compromise, or either house could reject the compromise. After continued behind-the-scenes work, on September 27 the House agreed to the Senate's changes, and the conference committee was avoided. The bill was sent to President George Bush.

The bill was still in jeopardy, however, because of the Department of Justice's continuing opposition. Justice department officials who had been appointed by President Bush, and who had steadfastly opposed the bill, could recommend that Bush veto it. In exchange for softening their opposition, Udall said, the department was granted its request to administer the program. This move would put the U.S. Department of Justice, which had consistently sought the defeat of the bill, in control of judging who would and would not be able to receive payments. At the urging of a host of western Republican senators, including Senators Hatch and Domenici, as well as others, President Bush signed the Radiation Exposure Compensation Act (RECA), Public Law 101–426, aboard Air Force One on October 15, 1990.

"Bush signed it much to my pleasure and amazement," Udall said later. After a dozen years of spending most of his time on the case with little or no compensation, Udall and the Navajos, the white uranium miners, the downwinders, and the military veterans of nuclear weapons tests had finally won some justice. Unique to the bill was an apology from Congress. "The Congress apologizes on behalf of the Nation to the individuals...and their families for the hardships they have endured."[54] Udall insisted that this be part of the bill on a moral and ethical basis. A similar apology had been included in the compensation act approved for the Japanese who had been imprisoned in the United States. "These people are entitled to an apology," Udall said. "These are the only two instances where the Congress and the President apologized to a group of people who have been damaged."

Justice apparently had been served, but unfortunately the fight was not over.

8 Healing the Earth

Perry H. Charley sits at his desk in a modest, paneled office inside a brown, double-wide trailer in Shiprock. The trailer rests near the site of a former uranium processing mill. Geologic studies, hydrologic and mining documents, and health impact studies spill out of three-ring binders that fill the shelves and cabinets of Charley's office. On his wall are maps and pictures, one of which is an enlarged blueprint of uranium mining claims in the far northeastern corner of Arizona on Navajo tribal lands. On a laminated map of the region, Charley has isolated and identified hundreds of abandoned uranium mines that he plans to fill, cover up, collapse, or otherwise seal forever. His mission is monumental: to undo the damage that has been done to the land and subsequently to the Navajo people by the nearly 1,200 mines that were dug on the reservation. He intends to heal the earth.

It is uphill work. Charley's Navajo Abandoned Mine Lands Reclamation office has a budget limited by the restrictions of the Surface Mining Control and Reclamation Act of 1977. With a small and dedicated staff, he is reclaiming the mines scattered in remote locations over the sprawling Navajo lands. Charley is motivated. His father, Harris Charley, was a uranium miner for twenty-eight years. As a child, Perry moved with his family from mine to mine on the Western Slope of Colorado and southern Utah. Today he lives in a house nestled among the piñon trees in the foothills of the Carrizo Mountains that overlook the small Navajo community of Beclabito, fifteen miles from Shiprock near the Arizona state line. Charley lives very close to the spot where he was born forty years ago. His voice became quiet when he explained that his father was buried nearby. His father began a long career in the mines when some of the first uranium mines located on Indian lands were dug not far from what is now his grave. Charley knows the uranium killed his father. "That's how I got started in this business," Charley said. "I wanted to find out what killed him."

From Charley's office one can see a large, sloped pile of round, gray rocks. It is what remains of the Kerr-McGee mill where crude uranium ore was brought to be processed into yellowcake. The yellowcake was then shipped out of state to other plants in the U.S. nuclear production complex, where it was further refined and converted into high-grade uranium used for bombs or fuel for energy.

The leftovers from the conversion process, the mill tailings, contain many long-lived nuclides, including thorium-230 and the highly toxic radium-226, which has a half-life of 1,630 years. Since thorium has a half-life of 77,000 years, it will produce radium-226 far into the future. Radium-226 is relatively soluble, and so leaching into groundwater supplies is a great concern.

The tailings piles, scattered across the western United States in twenty-four locations, are the responsibility of the Uranium Mill Tailings Remedial Action Project (known as UMTRA), an agency of the Department of Energy. In 1978, Congress passed the Uranium Mill Tailings Radiation Control Act, which ordered the cleanup of all uranium mill sites because of the potential health and environmental problems caused by radioactive contamination. The act was passed in the wake of public outcry against millions of tons of mill tailings that had been abandoned by processing companies and left exposed and blowing in the wind since the 1960s. In Grand Junction and Durango, Colorado, for example, these piles were almost in the middle of town. On the Navajo Reservation, the Shiprock pile was near the heart of town. In Mexican Hat, Utah, and in the small Navajo community of Cane Valley, Arizona, a generation had grown up living beside tailings piles.

The U.S. Department of Energy admits that as many as 5,000 separate locations nationwide, such as schools, homes, and public and private buildings, have been contaminated by the tailings, either from dust blowing off the piles, proximity to large amounts of gamma radiation, or use of the sandlike tailings in concrete for slabs and footings. Although much of the uranium has been removed from the tailings, as much as 85 percent of the original radioactivity remains from residual radioactive materials.

In Shiprock alone, fifteen different locations, called "vicinity properties," were contaminated by the tailings.[1] Small amounts of radioactive materials and heavy metals continue to find their way into groundwater adjacent to the banks of the San Juan River, which flows just several hundred yards away from the tailings pile. Downstream from the tailings site, a very limited amount of the water is used for irrigation or domestic purposes. The town of Shiprock pipes in its water from a separate source upstream. The San Juan is also less than a mile from the Mexican Hat pile, which is further downstream and not far from where the river flows into Lake Powell.

In its public explanation, the U.S. Department of Energy describes the problem of leaching in typical bureaucratese: "Groundwater beneath the disposal cell [tailings pile] and in the floodplain north of the cell at the Shiprock site is contaminated with arsenic, lead, nitrate, selenium and uranium. The contaminated groundwater beneath the cell is not widespread and is not used as a source of domestic or agricultural water. Although the limited contamination from milling activities is not a current threat to human health, future beneficial uses of the groundwater must be protected."[2]

What this carefully worded description does not say is perhaps more important than what it does. The contamination is not "widespread," but the Department of Energy does not say how far it is spread. The contamination is not a "current" threat, but again the agency does not say when and how it might become one. Answers to these questions are part of a study that DOE is doing on all mill sites, information on which it is reluctant to release.

Originally in two piles, the Shiprock mill tailings were consolidated, covered with about seven feet of impermeable soil sloped so that water will run off, and then covered with about three feet of rock. The Department of Energy is quick to note that about forty Navajo construction workers were hired and about $1.5 million was paid in wages for the duration of the project, from July 1985 to February 1986. Despite the work, the groundwater around the Shiprock mill tailings pile remains contaminated by radioactivity and toxic heavy metals. A series of wells on the banks of the San Juan are being used to monitor the potential for discharge of radioactivity into the river, which so far, DOE spokesmen say, has not occurred.

At a public hearing in Shiprock on February 17, 1993, Don Metzler, an official with the Department of Energy's UMTRA project in Albuquerque, explained the situation to a handful of Navajos: "We're going to continue to monitor it [the groundwater] to make sure no one is harmed or endangered.... Some [radioactive] particles leached out of the tailings and into the soil.... The particles move very slowly with the groundwater to the terrace." The terrace Metzler referred to is a broad, flat area below the tailings pile and on the banks of the San Juan River. "We do have some contaminants down in the flood plain [terrace]," Metzler said. The water in the floodplain, he repeated, "is definitely contaminated." But, he assured the small group, as the contamination reaches the river, it is instantly diluted and washed downstream.

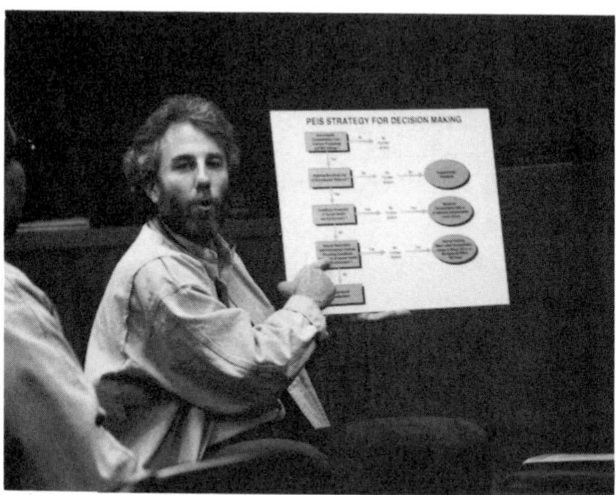

DON METZLER OF THE DEPARTMENT OF ENERGY ATTEMPTS TO EXPLAIN HOW THE DEPARTMENT MAKES A DECISION ABOUT CONTAMINATION OF UNDERGROUND WATER SUPPLIES NEAR URANIUM MILL TAILINGS PILES AT SHIPROCK, NEW MEXICO.
PHOTO: PETER EICH-STAEDT

PERRY CHARLEY, NAVAJO URANIUM MINE RECLAMATION MANAGER, WALKS AWAY FROM ONE OF MANY OPEN URANIUM MINES ON THE NAVAJO RESERVATION.
PHOTO: MURRAE HAYNES

"We pretty much have a good handle on our groundwater conditions at all our sites," Metzler said. As the DOE monitors the water, it will take action if the contamination appears to be spreading, he said. If the amount of contamination is low and will be naturally "flushed" out in within 100 years, the DOE probably won't act, he said. If the problem is a threat to the community,

Metzler said, the water could be pumped out, treated to remove the contamination, then returned to the soil. "There are all kinds of innovative ways to treat water in the soil," he added.

Contamination of the area surrounding the former mill is nothing new. The Kerr-McGee mill in Shiprock was built in 1953 to process ore from the company's uranium-rich Cove and Mesa mines in the Carrizo and Lukachukai mountain ranges across the Arizona border. Kerr-McGee operated the mill for ten years, when it was sold to the Vanadium Corporation of America, which operated it until 1968, not long after VCA merged with Foote Mineral Company. Over the life of the mill, about 1.5 million tons of ore were processed, creating a pile of radioactive sand that covered seventy-two acres.[3]

During demolition of the mill in the late 1970s, widespread contamination from yellowcake was found. About $100,000 worth of rich uranium oxide was found between two layers of plywood that had once been part of the roof of the mill. In 1979, in an impassioned appeal for assistance to Sen. Pete Domenici, Harold Tso, the former director of the Navajo Environmental Protection Commission, described what the tribe had found and its awful implications.

> When this roof became weathered, it was replaced with a second similar cover of plywood that was placed over the original roof. The second roof was covered with tar paper and gravel. The entire roof was adjacent to the mill roaster whereby uranium precipitate was oven-dried. We believe that the uranium dust was vented out toward and deposited on the roof. It appears that the U.S. Atomic Energy Commission who performed required and periodic radiological inspections were not cognizant of and did not include the likelihood of this hazard in the inspection regimen....The question is: What about the workers within and without the mill facilities who relied on this milling company and USAEC for health and safety?[4]

* * *

In mid-February 1993, Perry Charley stopped his truck in a lonely valley about five miles off U.S. 160 on the northern edge of the Carrizo Mountains in a corner of the reservation called Tse-Tah. He turned on a small device called a scintillometer, an updated version of the old Geiger counter used to measure radiation. The device screamed. He smiled. The truck was parked beside a large pile of gray-brown rocks and tan dirt covered with weeds and grass. On the other side of the truck was a similar pile of gray rock mixed with red dirt. These are piles of the waste that was extracted from several small uranium mines dug in a rocky depression several hundred feet away. The mines are long gone now, covered by Charley's crew with red dirt that has been seeded and covered with straw. This is a typical Navajo reclamation project, Charley explained. The small complex of mines was analyzed for relative danger from collapse, readings were taken, and then he and his crew figured out the best way to fix the mines. In many cases, because the waste was very radioactive, it was bulldozed back into the mines as much as possible, then covered with uncontaminated dirt, recontoured, and seeded. Where backfilling was impossible, the mine shafts and openings were covered with slabs of concrete or literally "plugged" with concrete, Charley explained. If the mine opening was badly deteriorated, it was blasted shut with dynamite. In the case of these small mines, the greatest threats to health have been solved, he said, because in many instances the old uranium mines were used as animal pens.

Although this small group of abandoned uranium mines may have been sealed and covered, Charley explained that there are still numerous small piles of radioactive uranium mine waste to take care of. Even inside the truck, Charley said, there is the kind of radiation a person would not want to hang around for any extended time. Fortunately, he explains, few people and few, if any, animals wander around here anymore.

Problems associated with these abandoned uranium mines and waste piles are complex and numerous. Navajo children and livestock play in or wander around abandoned mines. Unknown, unfenced, and uncovered shafts present dangers. Some of the mine openings look harmless enough but can lead to hundreds of feet of workings, Charley said, some of which may suddenly veer down or off to a side. The mines are also unstable, since most were dug out of sandstone rock, which fractures easily. The rock is great for short-term mining but highly dangerous in the long run because it will collapse. Many of the accidents and injuries that miners suffered came from mine walls that collapsed after drilling and explosions fractured the walls and ceilings. No Navajo children have been lost, injured, or killed by falling into an abandoned mine, Charley said, "so far."

The land is riddled with abandoned mines. Charley recalled that while he was working on one reclamation project, an abandoned borehole suddenly developed in the middle of a dirt road over which he and his crew were driving. It began as a hole measuring eighteen inches in diameter but large enough for a small child to fall into, and quickly expanded to expose old mine workings nearly sixty feet deep. It was backfilled and sealed with cement.

Falling into a hole is one thing, but there is always the unseen danger of radiation exposure, not only to humans but to animals. Of the 876 mines that Charley and his crew discovered during two years of combing just the eastern third of the Navajo Reservation, using old AEC production records and maps as their guide, they found that 467 abandoned mines were being used by animals for temporary or semipermanent shelters. In some cases, a rough fence around the opening of the mines made a convenient and natural corral and shelter that provided protection from the weather. Part of Charley's job has been to spend time with the local folks, explaining the dangers of such makeshift shelters—to themselves and the animals. Cows and sheep kept in such places become contaminated by breathing the residual radon, drinking some of the water that occasionally flows from or puddles in the mines, and eating the forage growing in earth that is contaminated with uranium waste and toxic metals. The contamination permeates the animals' flesh and bones, including the cows' and goats' milk, and then is consumed by local ranchers.

"There is just no way to keep it [radioactive contamination] out of the human population," he said. When he finds abandoned mines being used this way, he then must confront the rancher and find a way to explain the problem.

"How do you explain to them about alpha and beta particles, about gamma radiation or radon gas?" he asks. "There is no word for it." Charley likes to use the analogy of steam to describe radiation, but it is inadequate. "I call it steam, but they associate that with the ceremonial sweat baths, which are good."

Of the mines he has inventoried, forty-three were found to have been used by people for shelter during bad weather, perhaps on a regular basis. He found campfires, potato chip bags, soda cans, and food wrappers, he said. Some of the mines may have been used for hunting camps. In these instances, clothes and sleeping bags, food and water became impregnated with uranium dust and dirt. The radon-glutted air of the old mines was breathed. It is impossible to know how many people have been exposed or how heavy the exposure was, he said.

After the 876 mines were identified—a time-consuming and tedious job that took six men two years to complete—the worst mine sites were dealt with first. "We were interested in the individual mine sites [not just the lease claims documented by the DOE] and the physical hazards. We documented the inherent safety problems. We looked at usage, accessibility, proximity to residents, radiological hazards, and other extreme physical hazards. We looked at drainage patterns to see if radionuclides may be introduced into the water tables." The crew methodically eradicated the danger, leaving a fully covered or inaccessible mine.

However, determining what needs to be done is just the first step in the process for Charley. He departs from all similar reclamation projects by taking into account the social, religious, and ecological aspects of what he does. Accompanied by a Navajo biological scientist, Charley examines the wildlife and plant life affected. Since many plants living in the fragile and hostile desert environment do most of their growing and pollination in the spring, Charley said he and his crew normally do not do much of their construction and earth-moving work during that time of year. They often wait until the fall, when their work will do the least harm to plants and wildlife.

Raptors such as eagles, he said, nest in the spring and raise their young in the early summer. If any raptor nests are found, they are left alone until the fall or may be covered by a net during the winter so that when the raptor arrives in the spring, it is forced to find an alternative nest for that year. Once the reclamation work is completed, the protective net is removed, and the next year the raptor can use the nest again.

Charley's careful reclamation efforts do not end there. The reclamation areas are surveyed by archaeologists for ancient remains. If any are found, they are generally left untouched. If they must be covered because of contamination, they are thoroughly inventoried and documented—a time-consuming and involved process.

Charley said he also visits with the local medicine men to find out if his work might affect any spiritually sacred sites, sites used for traditional ceremonies and prayer offerings, or to gather medicinal herbs. If it will, he consults with the medicine men to determine how best to approach the cleanup. Such thoroughness and concern in reclamation work are generally unheard of. Charley and the Navajo Nation are ahead of the rest of the world and the U.S. government in this careful methodology.

Like the spread of nuclear weapons and nuclear energy itself, which has permeated our society and our thinking, the effects of the uranium mines have reached deep into Navajo life. From the top of a pile of radioactive mine waste, Charley gestured to a house nearby where he found that the traditional beehive-shaped bread oven had been built with radioactive rock from the abandoned mine that he had reclaimed. The oven was radioactive. "We put a stop to it [being used]," Charley said.

A few minutes later Charley again stopped his truck in the middle of an expanse of dry grassland. In the distance the snow-covered Ute Mountains rose into a deep blue sky. He gestured toward a house some way off with a traditional round hogan beside it. "There used to be two hogans there," he said. One was knocked down after he found out it had been built of uranium mine waste. However, the owner, although agreeing to knock it down, simply took the radioactive rock and wood to the nearest sandy gully and dumped it. The rock should have been hauled to a mine site and buried, Charley said. Instead, the radioactivity that endangered the owner is now being spread further and further out into the environment. Each time it rains, more and more radioactive elements are washed into the sand and carried downstream. Charley sighed, shrugged his shoulders, climbed silently back into his truck, and drove home.

* * *

Ray Tsingine, a tall, athletic Navajo, casually steered the white, four-wheel drive pickup off the side of U.S. 89 on the western edge of the Navajo Reservation in Arizona, sixty miles from the Grand Canyon. It was mid-March, and already the area was busy with tourists—buses, lumbering RVs, and station wagons filled with people looking at the seemingly dull and barren landscape, wondering how much further it was to the next gas station.

He stepped easily over a barbed-wire fence twenty yards from the road and walked over familiar dirt and rocks to a large pit. He stopped suddenly to pick up a rock and said, "Petrified wood. All of this gravel is petrified wood." The

MANY OPEN-PIT URANIUM MINES HAVE FILLED WITH WATER AND BECOME WATERING HOLES FOR ANIMALS.
PHOTO: PETER EICHSTAEDT

ground was covered with it, from pea-sized to baseball-sized chunks. He lightly hefted a chunk, knowing that in his cupped hand he held a rock that represented a piece of geologic time so immense that it is incomprehensible to the human mind. He let the rock drop with a thump and stopped at the edge of the open pit. The pit was sloped at one end so that a truck could negotiate it easily. At the bottom was a puddle of brackish, green water surrounded by a few scraggly bushes. On a long list of more than a hundred such open-pit uranium mines that dot the Cameron–Tuba City area, this pit bears the name of Jeepster. Most of the other abandoned uranium mines in the area carry the names of the former owners: Earl Huskon No. 1, Earl Huskon No. 21, Lemuel Littleman No. 1 and No. 7, Max Johnson No. 20, Julius Chee No. 3 and No. 4. Some bear more colorful names such as Montezuma No. 7C or Jackpot No. 40, reflecting a prospector's eternal optimism.

As with Perry Charley, his counterpart on the other side of the reservation, Tsingine has grouped the open-pit mines into three categories according to the degree of hazard they pose. The worst of them are being filled first. Jeepster will be filled soon, he said, "The water table is high here, and we don't want to take any chances." Tsingine is well aware that it may be too late. Some of these mines have been open and exposed for thirty years or more. Rain and snow have washed into the pits and leached heavy metals and radioactivity into the underground water table. Tsingine hopes that the radioactivity is at least sufficiently diluted by the time it reaches the water table and eventual human consumption that it will pose very little "immediate" threat to humans who might consume it.

However, the mines are by no means safe. Tsingine's concern has been for the livestock and wildlife that roam the barren landscape. In the hot, dry desert, any small pit where water gathers can be a life-saving oasis. Coyotes, foxes, cattle, horses, sheep, rodents, reptiles, and birds all drink from any water hole they find, instinctually knowing that it could be their last swallow for days. The foul liquid that gathers at the bottom of such old open-pit mines means sickness and even death for the unsuspecting animals.

The plan, Tsingine explained, is to first line such holes with a thick layer of inert material such as clay to provide an impermeable barrier between the water table and the toxic uranium waste. The hottest of the uranium waste is then piled on top of the clay layer, followed by the moderately hot and finally the least hot. The final layer is the old, inert soil, called "overburden," that originally covered the uranium deposit before it was uncovered and excavated. The fill, much like the fill for the mines in the northeastern corner of the reservation, will come from the mounds where it was piled more than thirty years ago when it was first dug out of the earth. These innocent-looking dirt piles also pose a threat to all life forms that come near. Not surprisingly, little grows on such piles. These uranium waste piles give off radon gas and emit the dangerous and deeply penetrating gamma radiation. Repeated and prolonged exposure is a danger not only for wildlife and domestic stock but for the humans who own and harvest the animals. Since the animals are closer to the ground, they are exposed to higher levels of radiation than humans.

Prolonged exposure to gamma radiation can result in genetic defects that can be as subtle as general ill health or as severe as physical deformities. There has been no analysis of the long-term effects of exposure to such waste piles on animals or humans. Causes of health problems can be difficult to detect at best and often impossible to nail down. Even though the precise health effects of exposure to these piles have never been determined and most likely never will be determined, concerned people like Charley and Tsingine operate largely on instinct and common sense that tell them that prolonged exposure is not good for anyone or anything. When the job is done, Tsingine explained, the land will have been restored to a more benign condition.

A long, low, rocky plateau called Ward Terrace stretches south from the community of Cameron, Arizona. The focus of the plateau is a fortresslike, turn-of-the-century trading post made of dark sandstone, mortar, and painted old logs. The parking lot is filled with buses of elderly tourists who ride a circuit created by tour guides in Las Vegas, Nevada, and this is a lunch stop. About 500 yards from the trading post, the sandstone cliffs descend to the Little Colorado River, a muddy and meandering stream that begins a couple hundred miles to the east on Navajo land and empties into the Colorado River at the eastern edge of Grand Canyon National Park. The banks of the Little Colorado south of Cameron are an ancient alluvial fill of deep sand and rock. It was here that rich uranium deposits were found near the surface by the prospectors of the late 1940s and early 1950s. It was easy pickings. The dirt that covered the uranium was scraped away by bulldozers, along with the low-grade ore, and left in big mounds. The marketable uranium ore was scooped up by hand or heavy equipment and hauled to the uranium processing mill at Tuba City, thirty miles away.

Tsingine scrambled up the side of a deeply textured gray mound that overlooked the reclamation work on what was once a large open-pit mine, then paused at the top to scan the horizon. Below, huge pieces of dirty, yellow, heavy equipment belched black smoke and roared across an expanse the size of several football fields. The big belly-loading machinery lowered its scrapers and was pushed from behind by an equally monstrous bulldozer until the loader was filled. The loader's belly was closed, and it snaked across the flattened dirt, throwing dust and exhaust into the air and emptying its load as it rolled. A water truck quickly followed, dousing the dust.

Tsingine surveyed what was left of one of the worst mines under his jurisdiction: the Jack Huskon No. 3. Following the procedure he outlined, the mine was filled in at a cost of tens of thousands of dollars. Unlike the many burrow-like underground mines being reclaimed by Charley in the northeast section of the reservation, these gaping open-pit mines require the use of heavy equipment that can move thousands of tons of uranium mine waste and dirt in a short time. In all, 240,000 cubic yards have been scraped up, hauled, dumped, and leveled at the mine—enough to fill fifteen floors of an average building that covers a square city block.

Tsingine looked around and said that too many people lived nearby to leave the mines exposed any longer. "The closeness of the residences was one of the justifications for the program," he explained, as if any justification was needed. Ironically, this work covering the open pits had been criticized by the nearby residents, who became accustomed to letting their domestic stock drink from the old uranium mines. To quell their consternation, Tsingine said, the Navajo Nation built new watering holes and improved some old ones for the livestock. This involved a common ranching technique of using a bulldozer to pile dirt up across a natural low area, gully, or dry streambed, creating a dirt dam that would hold rainwater runoff for weeks at a time. The work does not fall within the strict guidelines of mine reclamation, Tsingine said, but is needed. "We justify it by getting the animals away from the abandoned mines," he said. "It alleviates some future health problems, and the people like it."

This mine site, as well as the one just a stone's throw away, called Juan Horse No. 4, will be filled and graded to approximately reproduce the slope and shape of the land before anyone began to dig in it. Although work has been completed on some sites and is progressing at others, much still needs to be done. One former mine, called the Yazzie 312, is now a small lake from which water is drawn to fill the trucks that are used to control dust at the other mine sites.

While the water in this small, man-made lake in the middle of the desert may be safe for reclamation work, that of another mine nearby is not. Called Max Johnson No. 9, this gaping open hole in the earth has become a favorite dumping ground for trash. Tsingine walked dejectedly to the edge of the old mine that he and his fellow reclamation workers have dubbed "the Dumpster." He kicked at the charred, black remains of an old tire and then at an empty bottle of Jack Daniels. "People come out here, burn tires, and get drunk." He sighed deeply and surveyed the mountain of rusted tin cans and busted bedsprings popping out of mangy mattresses. A couple of feral cats raced up the side of the mound and disappeared over the rim as old cans clattered to the bottom.

The silence was like a shout. This gouge in the earth, 300 miles from any major city, 20 miles from any town, is a microcosm of what modern America has meant to the Navajos. Like other people, some Native Americans were caught up in the Cold War scramble for riches to be gained from selling uranium ore found on their reservation. Navajos were lured from a rugged, pastoral life in relative harmony with nature by the promise of wages that would bring wood houses, cars, trucks, and appliances. After the cars and trucks were worn out, the dependence on them was not, so the Navajos sought jobs in cities, leaving their ancestral lands and traditional culture. For some, health was replaced by sickness. Left behind were open pits with radioactive water and toxic heavy metals. Sheep and cattle drank this water and ate the sparse grass and vegetation in the rocky, sandy soil. Now into this useless uranium pit were thrown tin cans that contained food grown a 1,000 miles away, paint and oil cans, trash and waste, the busted and broken, the ripped and torn, the material wealth that had been created by digging up elements from the land, processing them, refining them, tooling them, and making them into objects of comfort and desire. Into this pit everything was returning to the earth.

RAY TSINGINE, NAVAJO URANIUM MINE RECLAMATION MANAGER, STANDS AT THE EDGE OF AN OPEN-PIT URANIUM MINE THAT HAS BECOME A TRASH DUMP AND CONTAINS TOXIC CONTAMINATED WATER. THE DUMP IS LOCATED ON THE WESTERN EDGE OF THE NAVAJO RESERVATION.
PHOTO: PETER EICHSTAEDT

Tsingine looked over the edge and into the pit. The underground water table had risen because of the recent very wet winter and created a small puddle of greenish yellow water at the bottom of the pit. Contamination, radioactive and otherwise, was being leached into the underground water. That water could eventually find its way into the Little Colorado and shortly thereafter into the Colorado, which flows through the Grand Canyon and which is also the major source of irrigation water for much of the San Joaquin Valley, the largest source of fruit and vegetables in North America. Tsingine looked forward to filling this hole and burying what it was and what it signified, forever.

* * *

Terry Yazzie, about thirty years old and married, is the son of Luke Yazzie, the oldest resident of Cane Valley and the man who first brought the uranium hunters and officials of the Vanadium Corporation of America to deposits in the

NAVAJO MINER LUKE YAZZIE, SECOND FROM LEFT, STANDS BESIDE HIS WIFE, JUNE, AND HIS SON TERRY WITH HIS WIFE, VIOLET, AT THE EDGE OF THE MONUMENT NO. 2 URANIUM MINE TO WHICH HE LED GEOLOGISTS IN THE EARLY 1940S.
PHOTO: PETER EICHSTAEDT

hills behind his hogan. The VCA "discovery" became one of the biggest-producing and richest uranium mines on the Navajo Reservation. It was rich enough that VCA established a processing mill at the site. The mill began operating in the summer of 1955 and continued until 1967. When VCA pulled up stakes and dismantled the mill, the radioactive sand, as well as piles of unprocessed rock and an open uranium mine, was left behind.[5]

"We used to play in it," said Terry Yazzie of the enormous tailings pile beside his house. In a February 1993 interview, he said, "It was a good exercise hill. We would dig holes in it and bury ourselves in it." Terry's wife, Violet, said the hill was the first place Terry took her when she came to visit the Yazzie family. As the highest spot in the valley, that is where Luke Yazzie, a shepherd, would ride his horse every day to watch his small flock.

While Congress authorized the cleanup of the uranium mill tailings sites in 1978, it was not until 1990, twelve years later and twenty-two years after the mill closed down, that the U.S. Department of Energy went to the Cane Valley residents and told them that the pile was a problem. The Department of Energy then put a fence around the base of the tailings pile and every few hundred feet hung a small radiation warning sign on it.

Terry Yazzie, his wife, and children have lived with Luke just several hundred yards to the south of the tailings pile. Violet said, "The kids played up there in the summer. We sledded on that during the winter.... Nobody said nothing about it from the time they stopped working until the cleanup."

To the southeast of the tailings pile lives Ben Stanley, a nephew of Luke Yazzie. Stanley worked at the mill during the last five years of its operation and said the working conditions were horrible. Workers were given no protection from the thick dust in the mill. The pay was $1.25 an hour. The lack of basic protective equipment, such as steel-toed boots, resulted in Stanley's toes being crushed in an accident, and he is still bitter about the company's refusal to pay for his medical treatment. Stanley drove a truck that hauled crude ore from the mine to the mill. The trucks had no brakes or starters and had to be roll-started. Stanley's brother, Ned Yazzie, suffered a broken back when a truck fell on him, leaving him crippled for the rest of his life.

For most of the residents of Cane Valley, the opportunity for high-paying manual labor jobs in their backyards was appealing, despite the dangers. And when the mining company left, the big sand pile was an attractive source of building materials. As a result, like the cases in other parts of the reservation, most of the homes in the area were built with concrete made from the tailings, an ideal mix for cement. The mill water, Stanley remembered, was dumped into a settling pond from which all the valley livestock drank. Standing next to the tailings pile in early 1993, he said, "We butchered those animals and ate them."

Stanley is angry because he was not told until recently that the tailings are dangerous: "Now they say it is dangerous for the people to work here. What about the people who have lived out here all their lives? They ought to give them all a health exam to see if they're all right." Stanley said he has told his story to many people, including newspaper reporters and television film crews, but nothing ever happens. "I must be talking to myself," he said. "The reason no one ever bothered [with] us is that we're a small community." Despite the problem, Stanley said, poverty locks the residents in the valley. "We have nowhere else to go. No one to turn to. Everyone who has left has come down with cancer."

Although the residents of Cane Valley may not have been told of the possible health danger the tailings pile presented, the government was very much aware of it. In 1975, the U.S. Environmental Protection Agency became aware of possible radioactive contamination in the homes of the residents of Cane Valley.

> Sixteen of the 37 dwellings surveyed were found to have tailings and/or uranium ore used in their construction. Tailings were used in concrete floors, exterior stucco, mortar for stone footings, cement floor patchings, and inside as cement "plaster." Uranium ore was found in footings, walls, and in one fire place. Other structures, not used as dwellings, were also identified as having tailings and ore use.
>
> The reason for their use is that the residents have unrestricted access to the white sands of the tailings pile and the broken stones at the mine and former ore storage area. The most predominate use of tailings and ore was found to be in residential structures and has resulted in elevated radiation exposures to the occupants of these structures. Tailings were incorporated directly into the floor, foundation materials, and exterior stucco. This is not similar to their [tailings] use in dwellings in Grand Junction, Colorado, where they were used primarily as bedding materials for floors and foundations.[6]

The report stated that the highest exposures were from 332 milliroentgens per year to 402 milliroentgens per year, which is roughly triple the annual exposure of the average American. Radon levels measured in the residences at the time showed very low readings of about a 0.04 working level, but the report stated that the samples were taken during the summer when the doors and windows were open and the houses well ventilated. Readings taken during the winter when the houses were closed might have revealed much higher radon levels. In addition, the samples were not taken over several days but were so-called "grab" samples. The initial report concluded that longer term sampling was needed.

Regardless of the findings, the readings were not considered to be high enough to warrant any immediate action, and nothing was done. Some of the readings were attributed to the gamma radiation that was emanating from the mill tailings pile. The radiation and other environmental problems associated with the tailings were documented in a technical assessment of the Cane Valley site, referred to generally as the "Monument Valley site," dated March 31, 1977, by the engineering firm of Ford, Bacon, Davis Utah for the U.S. Energy Research and Development Administration. The firm took a number of radiation and radon readings in and around the pile and concluded that there was very little danger to the area residents in general. The radiation within one-tenth of a mile around the pile was within background levels, or that which is natural for the valley.

However, the study did not take into account the fact that most of the residents spent a lot of time *on* the pile, not a mile or less from it. The assessment stated that the greatest danger to the residents near the pile, such as the Yazzie family and Ben Stanley's family, was from breathing a high level of radon gas that continuously seeped out of the pile. Because there was more radon in the air, the problems that faced the Cane Valley residents were the same that faced the uranium miners. "Radon gas exhalation from the piles and the subsequent inhalation of radon daughters accounts for almost all of the total dose to people from the Monument Valley site under present conditions."[7]

RUSSEL EDGE OF THE DEPARTMENT OF ENERGY DESCRIBES THE CONSOLIDATION OF MILL TAILINGS ON A HILL OVERLOOKING MEXICAN HAT, UTAH, IN FEBRUARY 1993. PHOTO: PETER EICHSTAEDT

The report concluded that over a twenty-five-year period "The presence of the tailings can be expected to increase the rate of occurrence of lung cancer by 18 percent for those persons presently within 1 mile of the pile." However, the report also concluded that "As a result of the low population density, and the low radiation levels of the tailings at the Monument Valley site, the potential health impact of the piles is minimal."[8]

This conclusion is generally accurate as far as it goes but is based only on short-term measurements of radiation near the pile. It does not include the potential long-term effects on the children and families who live in dwellings that are contaminated with mill tailings. Children, whom doctors say are much more vulnerable to the effects of radiation than adults, spent a lot of time on or around the tailings pile, as did adults.

The report stated that the potential health problems caused by the tailings pile are magnified, not diminished, by the milling and concentrating process. The problems do not come from the uranium but from other remaining radioactive materials such as radium and thorium, which are concentrated by processing instead of being left dispersed and buried in nature. "About 85 percent of the total radioactivity originally in uranium ore remains after removal of the uranium. The principal environmental radiological impact and associated health effects arise from the ^{230}Th [thorium-230], ^{226}Ra [radium-226], ^{222}Rn [radon-222], a noble gas, and ^{222}Rn daughters contained in the uranium tailings. Although these radionuclides occur in nature, their concentrations in tailing material are several orders of magnitude greater than their average concentrations in the earth's crust."[9] As noted earlier, radium-226 is soluble in water and thus can be leached into the ground.

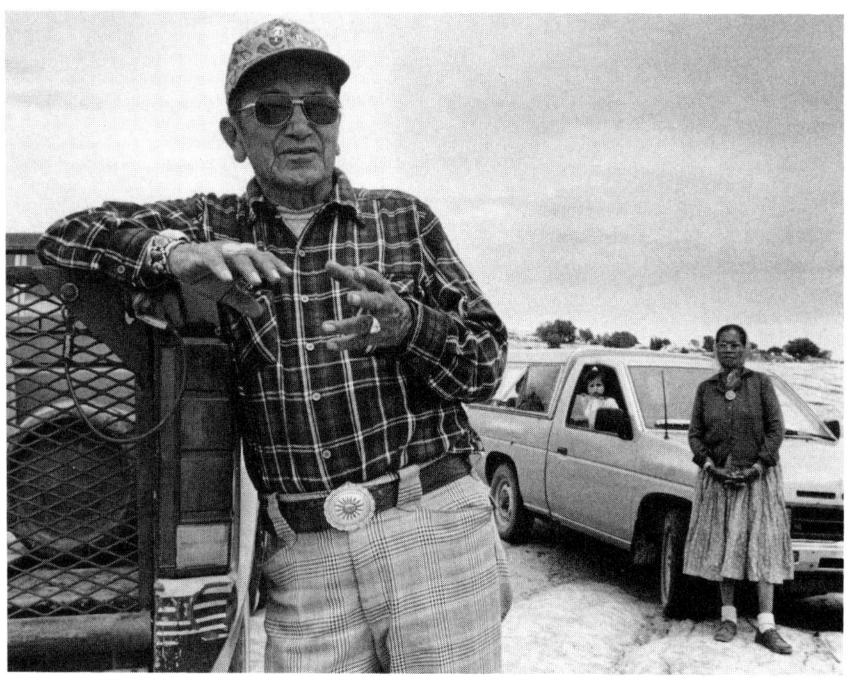

NAVAJO MEDICINE MAN ROGER HATHALE BELIEVES THAT URANIUM SHOULD NEVER HAVE BEEN MINED ON THE NAVAJO RESERVATION BECAUSE IT DISRUPTED THE HARMONY OF THE NATURAL WORLD.
PHOTO: MURRAE HAYNES

Despite the conclusions in this assessment report, in the late 1980s the U.S. Department of Energy decided to move the mill tailings pile in Cane Valley (in Monument Valley) to another pile at Mexican Hat, eighteen miles away. Supervised by Department of Energy employee Russel Edge, the site representative, the move began in 1992 with grading, widening, and improving the road between the Cane Valley pile and the Mexican Hat pile. The reason the tailings pile is being moved instead of being left and "stabilized" is somewhat nebulous. Despite the estimated cost of from $5 to $10 million, the move will save the DOE the costs of stabilizing one more pile. In an interview in early 1993, Edge explained it is also in the best interest of the residents of Cane Valley. When the tailings pile is finally removed, the land under the pile and surrounding it will be cleaned to premill conditions, he said.

In many ways, the consolidation project is unique. Edge, who had been with the DOE only a year when he took on the project, knew he was facing a lot of community resentment and distrust of the Department of Energy. It was no secret that many of the uranium miners had died of cancer, and deep-seated and justifiable fears of pollution from radioactivity were widespread in the community. Edge said he talked with leaders in the adjacent Navajo community of Halchita and found that the entire project and the Mexican Hat mill pile had been cursed by local medicine men. If he could do anything to begin rebuilding a bridge between the community and the project, Edge figured a spiritual leader might be a good place to start. He was told to get a medicine man to bless the project and that Roger Hathale would be a good one for the job. Hathale agreed and was paid $500 to perform a traditional Navajo blessing ceremony to launch the project.

In what Edge is certain was a first for the energy department, officials were flown out from Washington, D.C., as well as from the headquarters of the high-priced construction and technological firms that had contracted to do the project. Hathale's two-and-a-half-hour ceremony included having the stuffy bureaucrats and high-powered executives strip to their waists, dab their arms with corn pollen, and get down on their hands and knees to think good thoughts.

Hathale lives in a remote corner of the Navajo Reservation, about an hour's drive from Halchita. Using his daughter Mistena as an interpreter, in June 1993 he talked about the ceremony and uranium while standing on a rock outcropping not far from his house. Hathale said he performed the blessing ceremony because he was asked to do it. He explained that because of the past danger that uranium had been to people, there was a lot of hesitation about having any more dealings with it. "People said that they didn't want the uranium removed because it killed a lot of people," Hathale said.

However, the project to remove the tailings from Cane Valley was good. A deeply spiritual and traditional Navajo, Hathale said the problem with uranium goes back to disturbing Mother Earth. "He said it [uranium] should never have been dug up. He believes in Mother Nature," his daughter explained. "To him, everything that has been taken out of Mother Nature has been harmful." Hathale knew long ago that Luke Yazzie's father, Adakai, did not want Luke to show the white men where the uranium ore was. "Back then those old folks like his dad [knew] that it…is poison and can kill. They knew it, but it was the white men that took it out and it killed a lot of Navajos. It causes a lot of sickness."

Hathale said that the death and environmental destruction that uranium has caused is only one more example of the worldwide environmental problems that modern society has caused. "Mother Nature is getting angrier," he said. The health problems that are being suffered globally can be simply explained, he said. "It's because of what is being done to Mother Nature." Because the traditional Navajo view is one of nature and man living in a harmonious balance, when that balance is disturbed, such as by "digging through Mother Nature" to remove minerals, that harmony is disturbed, and death and disease result. To restore health, one must first restore the harmony.

Edge said Hathale was asked to perform a second blessing on the project once it was completed. Edge got more than the blessing of Hathale. Of the approximately 150 people who were eventually employed on the project, a majority of them were Navajos. By March 1993, nearly 100 Navajos had been employed and about 25 non-natives were working on the project, mostly in technical and managerial areas. Most of the Navajos were trained to be truck drivers to haul the tailings from Cane Valley to Mexican Hat and gained a marketable skill—commercial driving—by the end of the project. Others were trained as radiation and engineering technicians and as surveyors. Edge estimated that about $5 million in wages would be paid into the community before it was all over.

What will remain will be a large consolidated pile of about 2 million cubic yards of mill tailings and the scrapings from the area surrounding the Cane Valley pile. The Mexican Hat pile will cover sixty-five acres and will be perched on top of a bluff of red sandstone rock overlooking the San Juan River. The pile will be covered with two feet of clay, eight inches of gravel, and another six inches of rock. It will be shaped to shed water and is expected to stay in place for hundreds of years.

The energy department is eternally optimistic. As one views the wide expanse of the tailings piles, with tons of clay and rock covering them with a supposedly impermeable layer, one can only speculate about the eventual long-term effects of such piles. Upstream on the banks of the San Juan, the Shiprock tailings pile has already leached radioactive elements into the underground water beside the river.

Even with seven feet of clay, rock, and gravel covering the Shiprock pile, Perry Charley is among those who are doubtful that the tailings piles are truly "stabilized." He pointed out the pesky salt cedar bushes that have been able to sink taproots deep into the tailings pile. The plant roots penetrate the seemingly impenetrable barrier of rock and clay and crack it, allowing rain to enter the pile and continue the steady leaching of radioactive elements into the groundwater below the pile. The cracks also open pathways for radon gas to escape into the environment. Unless the tenacious and fast-growing plants are regularly pulled out, the pile is hardly more stabilized than if it had been left uncovered. As is the situation at the old Kerr-McGee mill in Shiprock, according to public pamphlets distributed by the Uranium Mill Tailings Remedial Action Project, radioactivity from the tailings pile at Tuba City is leaking into the groundwater that feeds the household and farming needs of the Hopi village of Moenkopi, just five miles away.[10]

One cannot help but think about how long this radioactive material will be around—essentially for hundreds of generations—and how eventually the tail-

THE JACKPILE MINE ON LAGUNA PUEBLO, WEST OF ALBUQUERQUE, IS BEING RECLAIMED BY A PUEBLO-OWNED COMPANY.
PHOTO: MURRAE HAYNES

ings will be washed into the river, even if it takes hundreds of years, and distributed along the course of the riverbed.

* * *

More than forty years after work first began on the sprawling open-pit Jackpile uranium mine at Laguna Pueblo, bulldozers and earth movers still crawl over the exposed earth. Instead of digging up uranium-bearing ore, they are filling in this gaping hole. James Olsen, Jr., the reclamation project manager, said the source of the mine was a unique geological occurrence that can be traced back to the Jurassic Age when the uranium was deposited in this sandy river bottom over millions of years. Standing on the bank of a pond of turquoise-green water at the deepest portion of the mine, Olsen described the reclamation plan. Some 600 feet above, the remains of the original landscape towered into the deep blue sky. As he talked, bulldozers pushed tons of pale sand into the deep, standing water, filling it with inert material.

Olsen has made a career of the Jackpile mine, taking a job as a mine manager in 1977. The mine closed in 1982 after thirty years of producing more than 24 million tons of some of the richest uranium ore ever found in the continental United States. It had an average of 2.5 percent uranium, yielding about five pounds of oxide per ton of ore. Olsen said the Jackpile ranked as one of the top four or five uranium producers in the world. When the company decided to make the mine an open pit, the mining was conducted by breaking up the sandstone with heavy equipment, scooping it up and hauling it to a crusher, then transporting it by open train cars to the Anaconda mill in Bluelake, near Grants. The mine produced about 6,000 tons a day and filled two trains a day.

From 1982 to 1989, the mine lay open and exposed until the reclamation project was launched by Laguna Pueblo and its Laguna Construction Company, with about $40 million from Atlantic Richfield and the Anaconda Mining Company, which merged in 1970. Although about 650 miners were employed at the peak of its operation, just 60 are used for the reclamation work.

The reclamation involves taking nonradioactive material and using it as filler; this is followed by low-level radioactive waste ore, then a layer of shale about one foot thick; the mine is covered by eighteen inches of topsoil. The topsoil is then seeded and planted. The land is expected to be ready for grazing within a year or two. Besides covering the radioactive material, the main concern is protecting the groundwater from contamination by radioactive materials that might leach into it, Olsen said during a May 1993 tour of the mine. The village of Paguate, which overlooks the mine, draws its water from a different underground formation, Olsen said, and it is therefore safe: "The Jackpile aquifer is not connected with the source of drinking water for Paguate Village," he said. "It's isolated from this with several hundred feet of impermeable material." The water at the bottom of the remains of the open-pit mine has radioactivity in it that exceeds standards.

Although Olsen is confident that the groundwater will not be contaminated by the mine, monitor wells have been set up downstream from the creek that runs through the mine site. "We don't get anything of any consequence," he said. As part of the reclamation program, water quality will be monitored for ten years after reclamation is complete. In addition, all other aspects of the project will be monitored, including air and land. The natural or background radiation was 14 microroentgens per hour, and the mine itself will be put back to 28 microroentgens per hour. Olsen said the shale layer that is put over the low-grade ore will achieve that basic standard, and with the addition of eighteen inches of topsoil, the land above the mine site will have only natural background levels.[11]

Olsen said that one of the ironies of the project is that there are no federal standards for the cleanup of uranium mines. The work at the Jackpile is unique in the world, he said. The cleanup standards that are being met are the result of several years of work during the mid-1980s by the U.S. Bureau of Indian Affairs, the U.S. Bureau of Land Management, Laguna Pueblo, and Arco-Anaconda.

While there are questions that surround the eventual cleanup of such mines as the Jackpile and the hundreds of mines on the Navajo Reservation, the fact that cleanup work is being done at all is unique. Despite the host of environmental laws such as the National Environmental Policy Act of 1969 and such massive cleanup programs as the Superfund, there exists no specific or organized plan to clean up uranium mines in the United States.

The Uranium Mill Tailings Radiation Control Act placed millions of dollars into the effort to clean up mill waste, but nothing was done to attack the problem of the uranium mines that dot the West. Native American communities have piloted their own programs to reclaim their lands. In the next decade, it may very well be that countries from around the world—uranium was mined in Africa, Asia, and Australia, as well as North America—will be visiting the Navajo Reservation and Laguna Pueblo to study the reclamation of uranium mines.

9 RECA Revisited

In February 1993, when Dr. Louise Abel of the Indian Health Service hospital in Shiprock concluded her report on the uranium miners, the rest of the tragedy was revealed. She had already shown poignantly and factually the deficiency of the medical tests required by the Department of Justice bureaucrats. These tests were based on those used to determine compensation for victims of black lung disease. Abel demonstrated to the assembled doctors, lawyers, and government officials that the medical tests were inadequate. Out of nearly 600 miners tested who had worked more than a year in underground mines, only 5 qualified for the compassion payments that Congress ordered through the Radiation Exposure Compensation Act in late 1990.

For example, the pulmonary tests used to gauge how much a miner's lungs are damaged were administered while the miner was seated. It was only when miners moved around, when they tried to do outdoor chores like gathering wood or feeding their animals, that they were short of breath. The breathing tests, a rough measure of lung capacity, were designed to identify victims of impaired lung capacity. A stress test, such as placing the miner on a treadmill, would be more effective. However, the hospital had no money for such equipment, and it would be very difficult to get elderly Navajos to climb on the device with a breathing apparatus attached to them and wires on their chests. Abel also said the kind of stress necessary to obtain accurate information might trigger a heart attack. Some of the miners had to be carried into the hospital for their examinations, she said. The pulmonary function test was not the best test to determine lung impairment, and "exercise testing is very expensive and other tests are invasive," Dr. Abel said, especially to Navajos already leery of modern medicine.

Dr. Abel said the results of the pulmonary test showed that less than 10 percent of the miners had restricted breathing. Of the 516 miners given the seated pulmonary test, only 8.3 percent, or 43, had a loss of 25 percent or more of their normal lung capacity. Forty-three percent of those tested had some lung impairment but not enough to qualify under the justice department's requirements.

The second test required by RECA is two examinations of chest X-rays by what are called "B readers," doctors or technicians trained to identify irregularities. Of the 264 miners who had chest X-rays, 31 had two positive "B" readings, which indicated lung diseases or impairments. Yet the majority of these 31 miners showed little or no impairment from the breathing test, which can happen with silicosis. Nine of the miners had only one positive "B" reading, which left medical people like Dr. Abel in a quandary. The IHS was already stretched to its financial limit, and Abel said it had no funds to pay for a third "B" reader so that the results of the X-rays could be clarified. These readings cost $18 to $50 each. "We counseled them [the miners] that it would be good to have another reading," Abel said, "but we just don't have the funding."

Of the 549 miners who received some sort of testing, only 5 had two positive B readings and a sufficiently reduced lung capacity to qualify for the compensation. "The number stands out as very low," Dr. Abel said—an understatement. "These resting [pulmonary function tests] are not the best tests," she complained. "There are other kinds of radiological evaluations" that are better than B readings of X-rays. The tests showed that 43 miners had suspicious results in one of the tests or the other, but not both. Abel suggested that the sophisticated but expensive CT (computed tomography) scans are better for establishing the presence of disease. Because establishing disability is complicated, and simple tests are not sufficient, Dr. Abel said, "These men deserve further evaluation. The Indian Health Service does not have the capability to do it."

Helene Goldberg of the justice department and the top administrator of the compensation program was in the audience listening to the testimony. In her own defense, she told the group that the regulations her department drafted for qualifications did not place any restriction on which tests could be used to demonstrate the medical problems. However, Goldberg missed the point. The practical result of such an approach is that if a person has access to good medical facilities and can afford them, there are any number of tests that could qualify a miner for the $100,000 compensation payments. The fact was, however, that this was the Navajo Reservation, where the average annual income is about half of what the government considers the poverty level. This was the Navajo Reservation where at least half of the people, and almost all of the elderly who could qualify for the payments, did not speak English or trust the white doctors or modern medicine. This was the Navajo Reservation and this was Shiprock, about 200 miles from Albuquerque and the nearest major city, some 300 miles from Salt Lake City and some 400 miles from Denver. This was the Indian Health Service, which was able to provide basic and emergency care only to the most urgent cases. That the screening project existed at all was a minor miracle. IHS doctors and administrators felt the miners' problem was so severe it needed their attention. Money for the screening and tests, as well as the staff to administer the tests, came at the expense of other medical needs at the Shiprock hospital. Medical supplies and personnel were diverted for the miners' project. Yet the response from the justice department was that there were other tests if people wanted to get them.

Among those who were outraged at the February meeting was Stewart Udall, who waited impatiently for a chance to speak. "There's a big difference between Denver and Dennehotso," Udall chided, referring to the small Navajo community about fifty miles south of Monument Valley. "Rather than take money from vaccination programs and the like we should look to some other source," he said. "It pains me to see this put a strain on your [IHS] programs that we all know need to be improved."

Udall was already well aware that lack of money for appropriate testing was just one stone in the wall of regulations concocted by a squad of hard-nosed government lawyers in the justice department. It took eighteen months after Bush signed the Radiation Exposure Compensation Act for compensation money to start flowing to the Navajo victims and their survivors. On May 27, 1992, in an emotional ceremony at Window Rock, Arizona, the capital of the Navajo Nation, several hundred Navajos and Navajo leaders, as well as Sens. Pete Domenici and Jeff Bingaman were on hand to witness the delivery of the first checks to four widows of Navajo uranium miners. One of the widows, Daisy Mae Begay, a plainly dressed, elderly woman, had to be helped up to the podium to get her check.[1]

However, six months later and long after the excitement of the ceremony, the flow of money to the Navajos slowed to a trickle. Hundreds of claims that had been filed by Udall and other lawyers on behalf of miners, their widows, and survivors were stuck in the bowels of the bureaucracy. It became clear to Udall after he talked with some of the other lawyers that checks were moving to non-Indians twice as fast and twice as often as they were to Indian claimants. In an open letter to the justice department which he also released to the press, Udall unloaded.

"You have grossly neglected the applications of elderly uranium miner widows who should have been accorded places at the head of the line," Udall wrote in a December 17, 1993, letter to Stuart M. Gerson, assistant attorney general, and Helene Goldberg. "The pattern of your payments reveals an anti-Indian bias," and "the regulations your staff devised concerning documentation of medical facts severely penalize Navajo applicants." Udall provided a list of seventeen widows who filed claims in April and June of 1992 but whose claims were still unprocessed after six months. Udall accused the department of punishing the widows: "Has this inaction been prompted by a desire to punish these widows because they were in the group which went to court and proved that agencies of the federal government had sacrificed the safety of their husbands for supposed 'national security' reasons?"[2]

Of the many regulations, one that was particularly galling to people like Udall and Dr. Abel was the requirement that all medical records had to be "certified." Dr. Abel said she and her staff were dumbfounded at this rarely used requirement. Throughout modern medical history, hospital records have been considered official and authentic. A doctor's signature on a medical record is considered unquestionably accurate and truthful. "We had never heard of records certification," Dr. Abel said. Health records for the Navajo uranium miners were scattered and occasionally spotty. When the RECA was approved and regulations established, Abel said that the Indian Health Service was flooded with requests for up to 15,000 records a year, which forced the financially strapped service to hire several clerks who did nothing but duplicate documents. Because these clerks were not medically trained, a lot of unnecessary and irrelevant documents were copied and prepared for certification.

The certification process immediately became a monumental and time-consuming barrier. The result was that IHS records, which are Public Health Service records and therefore also U.S. government records, had to be sent to Washington, D.C., to be certified. It was one bureaucracy certifying the records of another at the request of a third agency. It was also creating long delays and huge backlogs while widows and other survivors suffered. Udall was justifiably incensed because this unnecessary process was delaying some compensation requests six months or more.

Meanwhile, records from private hospitals were being secured and certified in less than a week for most non-Indian claimants, Udall said. The problem was compounded for the Navajos because this bureaucratic step had to be explained to the Navajo-speaking widows, who could care less about a maze of requirements issued by a Washington, D.C., office. "When they don't speak English and don't have a phone, you're at a tremendous disadvantage," Udall said at the meeting. "When the government won't accept other government documents, it is downright outrageous if not silly. If you're an Anglo, the law works very well. But if you're a Navajo, you have a long, long road."

Then, to add to the certification burden, the justice department refused to recognize traditional Navajo marriages because these marriages did not include filing a piece of paper with the local county or tribal government. Again, the cultural insensitivity of the justice department became evident. Tribal marriages had been conducted by native spiritual leaders before the white man arrived and were sanctified in ancient ceremonies involving medicine men—people, not paper. Holy words, the vows of men and women, the witness of friends and families, a lifetime of living and occasionally the bearing of a dozen children were not good enough for the U.S. Department of Justice. A piece of paper, the product of European civilization, had to be produced. In the remote corners of the high desert of the Navajo Nation, such things are referred to in the Navajo language as *belagaana dola-bachan*, which means "white man's bullshit."

Goldberg had informed Udall that the reason for the document certification was to prevent fraud.[3] At the February 1993 meeting in Shiprock, and despite numerous protests, she restated the department's unbending and inhumane position. "We feel strongly that all records need to be certified. It is an important protection."

The U.S. Department of Justice, located in the heart of one of the biggest bureaucracies in the world, assumes that no one can live without documenting important acts of their lives. The idea of life without records was an alien and unsettling concept to the lawyers, who were claiming that the most powerful government on the planet had to be "protected" from a few hundred poverty-stricken Navajos.

A month before the Shiprock meeting, Goldberg's boss, Stuart Gerson, wrote a scathing letter in which he railed at Udall for failing to follow the requirements that the department had developed. Gerson said that the delays were Udall's fault, not the department's, and that the department was not at all biased. "In sum, your allegations of bigotry, unfairness and delay are completely unfounded and mask serious shortcomings in your fulfilling the uncomplicated requirements of the law," Gerson wrote. Yet as Gerson attempted to explain his department's actions, he contradicted himself. Each of the cases is reduced to a number:

> 201-16-927 Yazzie—This is a uranium miner claim filed by you on June 9, 1992, on behalf of a person who *claims* to be the surviving spouse of the deceased miner. When you submitted the claim, you sent an uncertified copy of a document that appeared to be a marriage license. When we attempted to verify the marriage with the Navajo Office of Vital Statistics, that office informed us that they had no record of a valid marriage. As you know, the Navajo Office of Vital Statistics keeps records of all marriages recognized by the Navajo nation, and has a mechanism by which individuals may validate Tribal custom or other unrecorded marriages.[4]

The justice department assumed that because the Navajo Nation has an office of vital statistics it has on record *all* marriages—an assumption that is totally erroneous. Until recent decades, when births, marriages, and deaths

NAVAJO URANIUM MINERS GATHERED AT THE MEXICAN WATER CHAPTER HOUSE IN MAY 1993 TO DISCUSS INADEQUACIES OF THE RADIATION EXPOSURE COMPENSATION ACT.
PHOTO: PETER EICHSTAEDT

occurred, few people worried about whether the event was properly documented in the eyes of the Navajo Nation or the U.S. government. In these cases, most of the miners' marriages took place forty or fifty years earlier, when no one bothered with documentation because it was neither customary nor necessary. Navajos lived their lives free of the white man's bureaucracy.

In the Yazzie case, Udall persisted, however, even though the situation degenerated into lunacy. In a series of letters and exchanges over the next six months after the initial claim was filed, the justice department insisted on a "certified copy" of a marriage license. Finally, Udall sent the department the original marriage license, but still this was not good enough! "When we examined the document [Yazzie's original marriage license], we discovered that the opposite side of the document was blank, indicating that no marriage ceremony had occurred."[5] The license had been provided, but apparently someone had not bothered to fill out the exact time and place the ceremony had taken place. Under the eyes of God, family and friends, and the community—even the Navajo Nation—the couple had been married, but because a couple of blanks on one side of the document had not been filled out, the justice department slammed the door on this destitute widow. Then, in a gratuitous act of self-righteousness, Gerson proclaimed that his battalion of bureaucrats had once again "protected" the U.S. government. "This case is a classic illustration of the need for original or certified records. Had we not demanded that you produce original or certified copies of the record of marriage, we would have paid $100,000 to a person ineligible by law to receive the money."[6]

It was painfully clear by the February Shiprock meeting that drastic changes had to be made in either the administration of the law or in the law itself. With the election of a new and, it was hoped, more sympathetic Democratic administration the previous November, there was the possibility

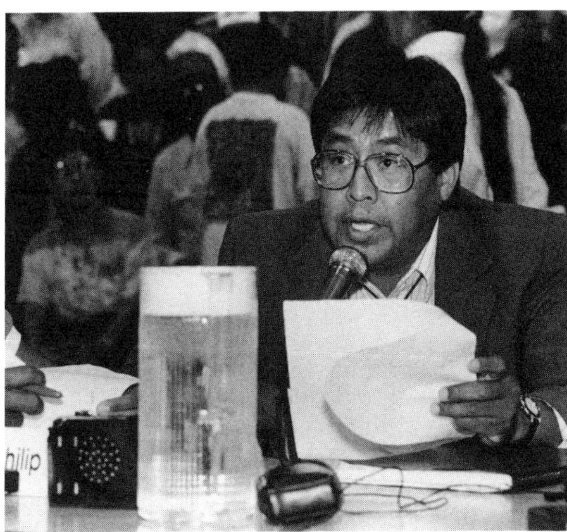

PHILLIP HARRISON, SON OF A DECEASED URANIUM MINER, HAS DEDICATED HIS LIFE TO HELPING NAVAJO VICTIMS AND THEIR SURVIVORS IN OBTAINING COMPENSATION FROM THE FEDERAL GOVERNMENT.
PHOTO: MURRAE HAYNES

that new people would bring a different attitude to the justice department. That, however, would be slow in coming. The alternative was to amend the compensation act or replace it with one so detailed that it would be impossible for any bureaucrat to subvert it. The congressional option was equally slow and probably more risky.

As they had many times before, the miners in the various chapters around the Navajo Nation gathered once again to list and organize their complaints. In March, April, and May 1993, they organized meetings and started work on a congressional hearing to be held in Shiprock. Witnesses and testimony for the hearing were organized by Perry Charley and Phillip Harrison, who, like Charley, was the son of a deceased uranium miner.

Harrison organized miners and millers, who were still not covered by the compensation act. Harrison, who formed the Navajo Uranium Radiation Victims Committee, and who has struggled to operate his own construction company while working to obtain compensation for miners and widows, has spent years of his life working on this problem. Frustrated at the slowness with which the justice department was issuing compensation claims to widows, Harrison flew to Washington, D.C., at his own expense to plead with the department to release funds. In several instances, he supplied explanations for some gaps in paperwork or provided the paperwork himself to complete the claims. In one visit alone, Harrison presented five hardship cases to the justice department, and as a result the claimants were compensated. He met on several occasions with department officials and agreed to act as a tribal liaison for the department to help ease the snarls of red tape.[7]

SEN. JEFF BINGAMAN OF NEW MEXICO, BACK TO CAMERA, LISTENS TO TESTIMONY FROM MAUDE TODACHEENE, FAR RIGHT, AND OTHERS DURING A CONGRESSIONAL FIELD HEARING IN SHIPROCK, NEW MEXICO, ON JUNE 5, 1993.
PHOTO: MURRAE HAYNES

Harrison was not used as a liaison at that time; however, he was reimbursed for his travel. Attorneys representing the Navajos are authorized by law to receive 10 percent of the claims that are awarded. This means that attorneys who are able to secure payments for their clients receive $10,000 each. Neither Harrison nor the Uranium Radiation Victims Committee has received any funds for their efforts.

Three years and three months after he conducted a hearing in Shiprock in 1990 which led to the eventual passage of the Radiation Exposure Compensation Act, Sen. Jeff Bingaman was back. In the same location, Bingaman and New Mexico Rep. Bill Richardson patiently and sympathetically listened to a litany of abuses that Navajo miners and survivors had suffered at the hands of the justice department. The bleachers of the gymnasium were filled with Navajos and more than a dozen news reporters and television cameramen.

Maude Todacheene, a sixty-eight-year-old widow of a uranium miner, slowly moved her frail frame to the speakers' table and addressed the group in Navajo. Her testimony was translated into English and entered into the hearing record. Like most other widows who had been left waiting, she said that she was frustrated in her attempts to secure a valid marriage license and the extensive medical records the department wanted on her husband, Frank, who had died in 1976. Since she and her husband were united in a traditional Navajo marriage ceremony, she thought that was all that mattered, she said, until she got a letter requesting validation of the marriage. Todacheene said she spent four months trying to get official certification of her marriage, which eventually was obtained from the Navajo Tribal Family Court. Todacheene, who was a client of Stewart Udall's, also complained about the difficulty of finding medical records for her husband, who had been dead for nearly twenty years, and then getting them certified.[8]

MAUDE TODACHEENE, WIDOW OF A URANIUM MINER, PROTESTED THAT HER TRADITIONAL NAVAJO MARRIAGE CEREMONY WAS NOT RECOGNIZED AS VALID BY THE DEPARTMENT OF JUSTICE.
PHOTO: MURRAE HAYNES

JESS WHITE, A FORMER URANIUM MINER, TYPIFIED MANY MINERS WHO WERE DENIED FEDERAL COMPENSATION BECAUSE THEY LACKED DOCUMENTATION, SUCH AS PAY STUBS, SHOWING THEY WORKED IN URANIUM MINES.
PHOTO: MURRAE HAYNES

Jess White, a miner who spent ten years in uranium mines in the Red Valley and Monument Valley areas, had been waiting for two years for his claim to be approved. White confronted another obstacle in the administration of the compensation act. This was the requirement to establish a verifiable work history. The justice department wanted such things as paycheck stubs or other tangible evidence that a miner worked in a certain mine during specific periods. The work dates are then checked against records kept by Dr. Victor Archer and others in the Public Health Service who had worked on the uranium miners' study. Work histories were also established by confirming that miners had been part of the PHS study. Although this valuable study was authoritative as far as it went, it was limited because it had not included all miners in all mines.

One of the concessions that the justice department won in the waning moments of the 1990 congressional session on the compensation act was that miners had to prove they had been exposed to at least 200 working level months in order to qualify for compensation. If a miner had been a smoker, then he had to demonstrate exposure to 500 working level months unless he had developed cancer or other respiratory disease before age forty-five. The rationale for this was the apparent synergistic effect of smoking and radiation exposure in creating lung disease.

Jess White's case was typical of the problems that the 200 WLM requirement created. Although he swears he worked for at least ten years in the mines, he is able to prove only a much shorter work history, which translates into only about 105 WLMs, far short of the required 200 WLMs. White testified about his problem: "I am not educated and do not maintain any of my old work history. I remember events associated with my mining experiences, but do not remember the dates, times and addresses of companies. Not being aware that my exposure would ultimately affect my health, I did not keep any records." White, like most of the other miners, worked in the mines for the money. As soon as he was paid, the money went for food, clothes, possibly a car, and gasoline. There was no reason to keep records because the miners rarely earned enough to pay income tax:

> As an uneducated man and maintaining a simple lifestyle in an extremely remote area of the Navajo Reservation, any trip away from my home requires a matter that is pressing. This usually entails my children verbally informing me of the situation. As I am unable to understand the writing on the letters I receive, I am unaware of their urgency until one of my children comes to visit.
>
> U.S. Department of Justice has to understand this [is] the typical lifestyle of many Navajos, especially the elderlies. We cannot afford to travel many miles to respond to every letter we receive. The lifestyle on our reservation is harsh and requires travel for many miles to obtain the necessities we need. These take precedence over many other things.[9]

White complained that the sixty-day limit to appeal and find new supporting information on such things as work history was simply unworkable for most Navajos.

It is more difficult to establish work history than even White and others realized. Udall submitted a set of letters to the hearing that detailed a case in which the justice department denied a claim by ignoring the work history and working level months that had been provided by Dr. Victor Archer. The case involved a widow, Mary J. Beyale, and her husband Keedah Nez, who had died of pulmonary fibrosis. A year after she filed the claim, the justice department turned Mrs. Beyale down on April 5, 1992, claiming that Nez had been exposed to only 41 WLMs and did not meet the 200 WLM requirement.[10] In his appeal of this denial, Udall included a letter from Dr. Archer explaining that Nez had been part of the uranium miners' study and had been examined in 1954, again in 1957, and then again in 1969. The mines in which he worked mostly were Mesa No. 3 and Mesa No. 2 in Red Valley. Mesa No. 3 had a working level of 24 WLs, and with all his exposure, Nez had a total of about 1,072 WLMs.[11]

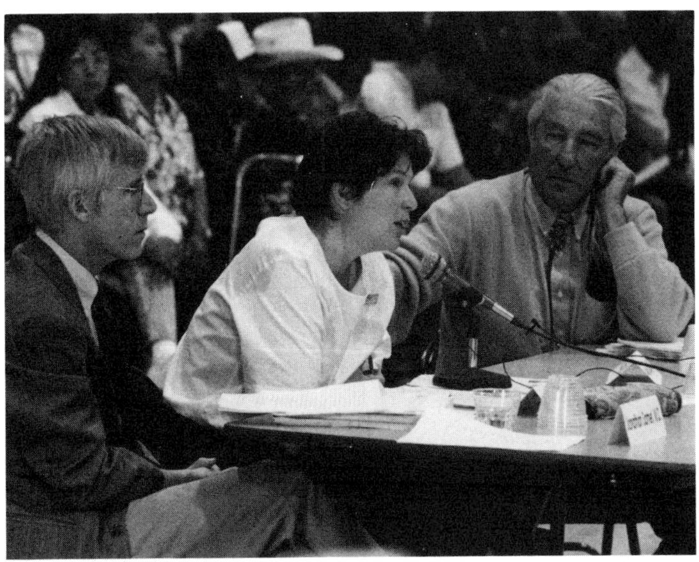

DR. JONATHAN SAMET, LEFT, DR. LOUISE ABEL, AND STEWART UDALL TESTIFIED TO THE INADEQUACIES OF THE RADIATION EXPOSURE COMPENSATION ACT AT THE JUNE 5, 1993, CONGRESSIONAL HEARING IN SHIPROCK, NEW MEXICO.
PHOTO: MURRAE HAYNES

After review, the claim by Nez's widow was finally approved, as were many of the claims that had been initially denied by the department. Why should someone have to go through all this to justify their claims and correct the numerous errors and assumptions of wrongdoing made by the Department of Justice? It is clear that without the help of attorneys who are accustomed to fighting such complicated paper wars with the federal government, the deserving Navajos would surely be uncompensated.

Equally objectionable is the distrustful approach the justice department has taken. In the Nez claim denial, Frank Krider, the assistant director of the torts branch which handles the compensation program, explained that the "preponderance of the evidence" has to prove the work history and exposure levels.[12] Such language demonstrates the intensely legalistic gauntlet through which a person has to pass in order to prove his or her claim as being worthy of the legal standard. In 1990, the U.S. Congress finally decided to distribute "compassionate" payments to people injured by the U.S. government. Yet obtaining these payments has somehow become a trial-like process in the hands of a battalion of stiff-necked lawyers.

It has been suggested that the program should never have been handed over to the justice department. Since these are payments for health problems created by mining, the program would be more logically administered by the U.S. Health, Education and Welfare Department, which has expertise in such issues. Another more logical department to administer the claims would be the U.S. Labor Department, which regulated working conditions in the mines and is well versed in workmen's compensation claims and related issues. In the hands of any other department, it would seem, the processing of the claims would not take on the mantle of the Grand Inquisition.

Udall made his points at a June 5, 1993, congressional hearing in Shiprock conducted by Senator Bingaman. "It is abundantly evident that Department of Justice officials who are administering the program are so obsessed with 'preventing fraud' that they are perpetrating an injustice by making it extremely difficult for Navajo widows and miners to qualify for payments under this program. Indeed, these officials' actions are nullifying the aim Congress had in mind when it extended a national apology to these citizens and set up a legal framework so they could be promptly compensated for the wrongs they suffered at the hands of their government."[13]

The justice department had overstepped its bounds, Udall said, when it misinterpreted the RECA, which intended compensation to go to victims of respiratory diseases by not allowing their attending physicians to be the authoritative source. The department, Udall noted, created the idea of two B readings for X-rays and a breathing test as the criteria, instead of a letter from the attending physician and accompanying medical records. The result of this bureaucratic maze was that of hundreds of Navajo workers who were potential claimants, only five qualified. One can only wonder why.

Also at issue is the working level standard itself. The RECA had originally been drafted with 100 WLMs as the qualifying standard. However, as the bill neared approval in Congress in 1990, Udall and others were fearful that it might be derailed at the last minute and acceded to Department of Justice demands that 200 WLMs be set as the minimum standard for nonsmoking miners and a whopping 500 WLMs for smokers. Udall argued that the standards should be drastically lowered. The reason was not so more miners could qualify, but because of the extreme difficulty of establishing work histories and obtaining accurate readings on the radiation and radon levels in the mines. In the early years of mining, when the mines were the "dirtiest," few readings were taken in the smaller mines, which were operated by small, independent contractors for just a few months or several years. "With these limitations in mind, it is clear that if justice is to be done, the WLM threshold should be lowered to 80 WLMs or less," Udall said. Udall argued that because a miner smoked, he should not be "punished" by being forced to prove he was exposed to much higher levels than nonsmokers.[14]

Udall proposed that the law be changed to allow attending physicians at the Indian Health Service to determine the presence of respiratory diseases for the Indians and to allow claimants to gather and submit information on their claims as they obtained it, not under arbitrarily imposed deadlines of the Department of Justice. He urged that the entire program be wrestled away from the grip of the justice department and transferred to some other more appropriate agency.

* * *

September 24, 1993, was the second day of a two-day meeting of the Southwest Indigenous Uranium Forum, an Indian activist group dedicated to stopping the further destruction of native people and lands by uranium mining and milling. The meeting was in the Laguna Pueblo village of Paguate, an ancient community on a mesa that overlooks the Jackpile mine. The meeting was largely the result of efforts by Alvino Waconda, a former miner in the Jackpile, who began to wonder about the effects that the mine had had on his village. Waconda asked the Laguna Tribal Council to endorse the meeting, but was refused. Waconda was able to obtain the endorsement of the village council, however.

The meeting was in the village hall, an ancient adobe building with hard-packed floors, painted walls, and a ceiling of varnished logs. The preceding day had been filled with testimony of days in the mines and the work experiences of many uranium miners from Laguna Pueblo and adjacent Acoma Pueblo. In attendance were a dozen or more environmental activists, some attorneys, a doctor and other experts from health fields, and a half-dozen curious but skeptical miners.

Manuel Pino, an Acoma Pueblo activist and teacher from Arizona State University, quietly took the microphone and began to talk about the day's discussion. The lasting impacts of uranium mining and the sorry record of RECA payments were on the agenda. Pino has written extensively on the hidden but lasting impacts that mining has had on the Pueblo people and their culture. His words were delivered in his typically soft-spoken manner but were pointed. "The economic incentives did not necessarily improve our way of life," he said. The justification for digging up dozens of square miles of pastoral Indian land had been profit and patriotism. There were millions to be made from the uranium, and besides, it was needed to stockpile weapons of mass destruction. Land was lost, but wages were gained; wages that would otherwise never have been earned; wages that went to buy appliances, cars, and all the rest that has come to represent modern America.

Pino, in his unassuming voice, quietly asked: At what price? As wages increased, so did what are euphemistically called "social problems." What had once been minor and infrequent problems became more widespread. Alcohol and drug abuse surfaced. Spouse and child abuse increased. "These incidents were parallel to the production of uranium," Pino said. The peak production years for the uranium mines were also peak years for suicides at the pueblo and

MANUEL PINO, AN ACOMA PUEBLO ACTIVIST AND TEACHER AT ARIZONA STATE UNIVERSITY, DISCUSSED THE SIGNIFICANT CULTURAL IMPACT THAT URANIUM MINING HAS HAD ON NATIVE AMERICAN COMMUNITIES.
PHOTO: MURRAE HAYNES

peak years for the dropout rate at nearby schools. It was "more lucrative to be a uranium miner and work at more than $10 per hour than it was to be in school," he said. Although the mining companies would argue that the pueblo residents were trained in a marketable skill, Pino countered that the skill was virtually useless once the mine shut down and the industry collapsed. The real result was unemployed miners with little or no education. In addition, a culture had been destroyed.

Such subtle and deeply insidious aftershocks of three decades of uranium mining have gone virtually unrecognized and undocumented by the engineers and lawyers who usually write environmental impact statements. What is most outrageous is that there has been little documentation on the health of the thousands of miners who worked in the Jackpile mine and other nearby uranium mines. The miners at Jackpile were largely ignored because the mine was open pit, and the dangers of breathing uncirculated air laden with radon were not present. However, that does not discount the dangers from breathing uranium-impregnated dust or the direct exposure to radiation emitted from the unearthed ore and its accompanying elements.

SUSAN DAWSON, PROFESSOR OF SOCIAL WORK AT UTAH STATE UNIVERSITY, SAID THE PSYCHOLOGICAL AND SOCIAL IMPACT OF URANIUM MINING ON NATIVE AMERICANS HAS BEEN ENORMOUS.
PHOTO: MURRAE HAYNES

"A lot of research has to be done," Pino said. As health effects are studied, the fact that traditional culture has been lost along with the language should be recognized. As miners bought cars and televisions, a once-foreign culture and language entered their homes. In just one generation, English replaced the Keresan language of the Laguna. Slowly but steadily, the traditional customs and ceremonies were replaced by non-Indian events. With clear outrage in his voice, Pino said that once the mining company asked if the Paguate residents might be willing to move their village because it sat above what was suspected to be the richest uranium in the area. "The traditional elders refused," Pino said. "This village is sacred land. You just don't move a village like that for economic incentives."

Pino later turned the meeting over to Susan Dawson, an outspoken professor of social work at Utah State University who has extensively studied the psychological and social impact of uranium mining on Native Americans, and who spoke at a 1990 hearing on RECA in Shiprock. She calls places like the Jackpile mine "technological disaster areas" with a lot of long-term social effects. For people like the Navajos, the Lagunas, and the Acomas, the mental trauma of finding out that they had unknowingly exposed themselves and their families to deadly elements was doubly severe because they have no word for or cultural concept of radiation.

Among the Navajos "there was a sense of betrayal," Dawson said, because the miners had been told they were helping the United States win the Cold War. "At no time were they ever informed of the risks of uranium mining," she said. The mining companies and Public Health Service officials who were studying the miners had agreed not to inform them. "The reason they did it was to insure the availability of a work force." The workers did not have a choice because "crucial information affecting their long-term health was withheld from them. There was negligence involved here," Dawson said.

Dawson is one of a handful of people studying long-term impacts on the people who worked in the uranium processing mills that dotted the western lands. Although there is anecdotal evidence and occasional scientific data that the millers may have been exposed to higher levels of radiation than the miners, their cases have never been documented, and the millers were excluded from compensation provided by RECA. This exclusion was justified during formulation of the bill because the conditions of the mills and the health of the millers had not been followed as closely nor had they been as well documented as those of the miners.

Ironically, it was the VCA mill in Uravan, Colorado, that aroused initial concern about radiation hazards. It was this mill that the AEC and Colorado officials inspected when they became concerned about working conditions on the Colorado Plateau in the early 1950s. In addition, it was the mills that first got the attention of Los Alamos scientists and others when questions about uranium mining safety were raised in 1949. Finally, it was millers and milling conditions that Duncan Holaday was asked to focus on during a 1957 PHS study.

Meanwhile, Dawson has interviewed dozens of millers to obtain their stories and to do a social-medical survey. Mills were places of intense exposure to uranium dust and highly toxic chemicals used to leach the uranium from the gritty ore. Ore from the mines was hauled to the mills and pulverized in open-air crushers. The uranium-bearing sand was saturated with chemicals that extracted the uranium, which after further processing, was sprayed onto hot drums. The excess liquids evaporated, leaving the drums coated with yellow uranium oxide which was, in another dusty procedure, packed into fifty-five-gallon drums and sealed for shipment. While miners were exposed to radon daughters and dust, millers were exposed to uranium and silica dust and concentrated uranium oxide. As can be imagined, millers frequently had uranium and other radioactive elements in their urine. This meant that the internal doses of radiation the millers were receiving were the potential danger.

Dawson's findings seem to bear this out. "The use of protective equipment [masks] was minimal at best," she said. "In the early days it was anything goes. Most workers didn't like wearing those masks. It was very uncomfortable."

Timothy Hugh-Benally said that by the end of September 1993, 329 Navajo millers had been registered by his office. In four separate chapter meetings in the vicinity of former mills, including Tuba City, Kayenta, Gallup, and Shiprock, former uranium mill workers gathered to formulate plans to obtain compensation and be brought under the wing of the RECA. "They unanimously said to sue the companies that created the situation on the reservation," Hugh-

Benally said. "They [the companies] abandoned their disasters and left them there for us to deal with."

Now the uranium mill workers, whose exposure may have exceeded that of the miners, will have to start down their own long road to obtain compensation.

Epilogue

BY THE SPRING OF 1994, the picture had brightened somewhat for the Navajo uranium miners and millers seeking compensation from the U.S. government for their losses. According to U.S. Department of Justice figures released in mid-March, 155 Navajo uranium workers or their families had been awarded the $100,000 compensation payments. This was less than half of the 324 claims that had been filed by that time, but it was an improvement over the number of claims that had been awarded just a year earlier.

In addition, 97 claims were awaiting a decision by the department. Seventy-two claims were denied for various reasons, the most frequent being that the miner was suffering from diseases not covered in the Radiation Exposure Compensation Act. Many of the claimants and their attorneys had appealed the denials, and twelve others were awaiting an appeal decision.

Apparently stung by Stewart Udall's criticism that white miners were getting approved for compensation twice as frequently as Navajos, the justice department evened the statistics on the award of claims. Of the total 1,567 claims filed nationally by former uranium miners and their beneficiaries, 743 had been approved for compensation. The approval rate now is 47.3 percent for white miners and 47.8 percent for Navajos. The rate of denials has also dropped sharply. Only 22 percent of the Navajo claims had been denied, compared with a 37.5 percent denial rate for miners nationally. The number of pending cases seems to indicate that Navajo claims still are more difficult to process, however. While 29 percent of the Navajo claims were pending, only 15 percent of the non-Indian claims were pending.

One of the glaring omissions of RECA was that workers in the uranium mills were left out of eligibility for compensation, despite documentation dating back to 1949 that they were exposed to hazards as severe or worse than those the uranium miners confronted. Although uranium millers were included in portions of the Public Health Service's uranium miners' study of the 1950s and 1960s, their specific health problems were not as extensively documented as those of the miners. Discussions about possible revisions to RECA at a June 5, 1993, hearing in Shiprock conducted by Sen. Jeff Bingaman resulted in a $500,000 appropriation for a study of the health and medical problems of mill workers.

The appropriation was sponsored by Bingaman and became part of the Department of Defense Appropriations Bill for 1994. The money was funneled to the army assistant secretary of health, who was given responsibility for developing guidelines for the study, putting it out for bid, and issuing a contract. The intent, of course, is to develop sufficient medical documentation to prove that mill workers were exposed to dangerous levels of radiation and suffered consequences similar to those of the uranium miners. As of March 1994, the request for proposals on the study had not been issued. Meanwhile, the Office of Navajo Uranium Workers was busy compiling a registry of living and deceased uranium mill workers. The list will be used to make it easier to collect data on mill workers and their medical histories in the pending study.

Timothy Hugh-Benally and Dr. Louise Abel said medical testing procedures for the uranium miners had improved from the previous year. Abel reported that exercise testing of miners had been arranged for the Navajos at the Miners' Colfax Medical Center, a state-run facility in Raton, New Mexico, that was built to treat coalfield workers. The testing is free for Navajos who are New Mexico residents, but those living in Arizona, Utah, and Colorado are charged about $200. A portion of the fee is covered by federal medical funds, but the remainder must be paid by the miners or their attorneys. Hugh-Benally explained that the Indian Health Service has secured a van and that eight to ten miners are taken from Shiprock to Raton once a month for testing. This presents some problems, he said, because most are very elderly and the eight-hour drive to Raton is taxing. Hotel and meal costs average about $100 a person for the trip. After the long drive, the miner is put on a treadmill the next morning, and after exercising, his blood is tested for oxygen levels. The tests are considerably more accurate than the seated breathing tests in determining lung impairment, Abel said. As a result, about 90 percent of the miners being tested this way are qualifying for compensation as victims of silicosis.

Although compensation is beginning to flow to the victims of uranium mining and their families, the bitterness and sense of injustice remain, as their words show. In March 1994 Perry Charley wrote, "My mother, Sarah, received her $100,000 check April 8, 1993...but my dad remains dead and I remain bitter...so continues the legacy of uranium miners."

Interviews

The following accounts are interviews with miners or their relatives conducted in February 1993 and April 1994.

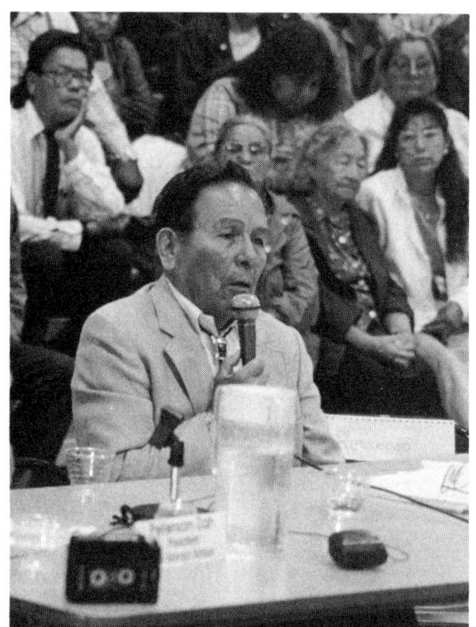
PHOTO: MURRAE HAYNES

TIMOTHY HUGH-BENALLY, age sixty-three, is director of the Office of Navajo Uranium Workers, Shiprock, New Mexico. Hugh-Benally directed a systematic search for living uranium miners who were in the general area of Shiprock, where much of the uranium mining on the Navajo Reservation took place in the 1940s, 1950s, and 1960s. Under a grant to the Public Health Service and the Indian Health Service, the search began in November 1992 and concluded in January 1993.

Hugh-Benally, who is now overseeing the Navajos' relatively extensive program to find, screen, and treat former uranium miners and millers, was himself a miner from 1956 to 1959. He was interviewed on February 10, 1993.

He recalls the years around 1950 when he was a teenager perched on the edge of a cliff and watched the mining operations in his home in the Oakspring area.

"In the late 1940s and early 1950s, people were all over the place looking for uranium. It was practically a stampede. Then they moved to the west and south to the Cove area.

"The first ones [mines] at Oakspring were Navajo owned and operated. But eventually they were turned over to other people for financial reasons.

"I worked in Cove with Kerr-McGee. I complained about the wages we were making, which were about 90¢ an hour when mining wages were $1.25 an hour. I got fired. They said there was no work for me. So I left."

He returned to uranium mining about 1962 and worked at the mines called Mesa No. 7 and No. 4.

"The working conditions were terrible. Inspectors looked at the vents. When they weren't inspected, they were left alone. Sometimes the machines [for ventilation] didn't work.

"They told miners to go in and get the ore shortly after the explosions when the smoke was thick and the timbers were not in place. There was always the danger of the ceiling coming down on them.

"The foremen were Anglos [white men] and rarely went into the mines. I complained about that. I said if we were represented by a union, we'd get our rights. I was fired again. I left it.

"The people I worked with are now all dead from cancer or other causes. Most of the rest are deceased.

"A lot [of those living] have filed claims, and some are still waiting. It's kinda sad to see people come and file claims and their claims are held up for various reasons. We filed some test cases and saw how they are processed. We'll see what happens.

"The people who worked in the mines started dying in the 1950s. We tried traditional ceremonies to cure the husbands. We tried traditional remedies, and some tried the Native American Church. They gradually went down. They were usually heavyset men and when they died, they were skin and bone.

"The U.S. government says it will compensate [us], but it is never enough. When they died, we had to carry on, chop wood and haul water. Now we have to go through strenuous records research.

"We provided uranium to the U.S. government, and [it] did not tell us it was dangerous. And now the government tells us you have to die to get compensated.

"Congress said it didn't know about the side effects of the uranium. It's ignorance on the part of the United States. They knew what the effects were. The people feel like they were a study project. The early studies prove they [the government] knew of the effects."

PHOTO: PETER EICHSTAEDT

JOHN BENALLY, age seventy-four, a resident of the Red Valley area, worked in various uranium mines for eighteen years, about seven of which were for Kerr-McGee in the Cove area and about ten of which were for various companies in the Colorado area. Benally started working in the late 1940s, about 1947, and continued until the 1960s. He has been hospitalized in the past. His comments were translated from Navajo on February 11, 1993, by Helena Benally.

"He had problems with his lungs and has never been hurt. He worked with a shovel and pushcart hauling out the broken rock after it was dynamited, a job called a mucker.

"He'd been telling us [there were] no vents in the mines. There used to be no air. Later they installed some pipes.... At night he feels bad."

Asked if he was warned of possible dangers in the mines from radioactivity, he said, "They never mentioned it." He was paid about $1.35 an hour or about $1.65 for mucking.

Another of the jobs was that of drilling holes in the rock in which to set the dynamite, a slightly more skilled job. "The ones that worked with the machines all passed away.

"He feels he was cheated by the government."

PHOTO: MURRAE HAYNES

KELLEWOOD YAZZIE, age sixty-seven, is a resident of the Red Valley area on the Navajo Reservation. He was interviewed on February 12, 1993, at the Red Valley Chapter House, and the translation from Navajo was provided by Lorraine Johnson.

Yazzie began working in uranium mines in the Cove area in 1958. He worked for about six years and was a drill operator. He also worked in the Slickrock, Colorado, area mines for about six years. He still has his paycheck stubs.

"He used to just substitute for people and operate drills and haul out the uranium. They used to dig out at least six feet and then blast. He used to only work in the mornings. He never really worked in the blasting."

The work was "dirty in [the] Cove [mines] but not in Colorado. In Cove they used to go way in, and there was not enough air to breathe.

"They never received anything to protect their mouths, and they were never told it would be hazardous. But then there were few jobs, and they had no choice but to work in the mines."

Wages were about $2.50 an hour, and after deductions were taken out, he took home about $160.00 for two weeks of work.

"They [are] all deceased, all the people he worked with.

"He has the pains, that pain going through his lungs and to the back. He thinks it may be from the uranium. He goes to get treatment, but they say there is nothing wrong. But the pain is still there. They told him there is nothing but high blood pressure."

PHOTO: PETER EICHSTAEDT

KEE BEGAY, age seventy-three, of Shiprock, worked in the mines in the Cove and Oakspring areas, and in Naturita, Colorado, beginning about 1959. His comments were translated from Navajo by Lorraine Johnson, on February 12, 1993, at the Red Valley Chapter House.

"When he was in Colorado, he was doing the blasting. One day the rocks collapsed on him, and he was temporarily paralyzed from the waist down." Begay was treated at the Indian Health Service hospital in Shiprock. "He was told he had another year [to live]," about 1961.

He sued the company he was working for and went to Denver to file a claim. "They did an examination on him and told him there was nothing wrong.

"Now he is on medication. They found uranium on his side." Begay pointed to his side to indicate where uranium grit was apparently embedded in his skin. "The pain is still there in his lung. He goes to Shiprock for it again. He's taking all kinds of medication." He has five different pills.

"Whenever he misses a dose, the pain comes back."

"He was never warned," about the dangers. "After two years he received a mask for protection from the dust. They were never told of the danger of uranium. At lunchtime they never washed their hands."

Begay had a son who died about ten years ago from some unknown disease. "The radiation was on him." He is convinced the family was contaminated because they all used the same basin for washing. "Back then there was no running water. They used a wash pan.

"He has been told his paperwork is going through, but he doesn't believe it."

PHOTO: MURRAE HAYNES

WILSON P. BENALLY, age sixty-five, of Red Valley, worked in uranium mines for twenty-eight years starting in 1948. He worked in Slickrock, Naturita, and Dove Creek, all in Colorado. His interview was translated from Navajo by Lorraine Johnson on February 12, 1993. Benally worked as a driller, mucker, and at other jobs. During the interview, he drew a map of one of the mines he worked in.

"It was not until 1963 that he was given one of those masks." He was also given a helmet with a lamp. Before that he used lamps that provided light from a slow-burning powder.

"When he first worked in the mines, he was given an X-ray and told he was OK. A mine collapsed on him, too," while he was working in Price, Utah.

Benally was recruited from the Oakspring area by a mining company out of Colorado. He was one of four hired.

"They were told they would use the uranium for cars, gas, and welding.

"When the rock collapsed on him, he almost lost his sight. And on one side, his arm was paralyzed. He was being treated in Albuquerque.

"He was told the uranium was a bit hazardous to the health, but he was never told how it might affect his health later.

"They used to eat underground and never washed their hands. They did some tests two years back, and they found dust in his lungs." Benally was part of the uranium miners' study, but "they never told him the results.

"So now the problem is that he [is] short of breath and sweats a lot, and he feels faint. He's under medication. He's using an inhaler now. They gave him some pill and told him to put it under his tongue; if it doesn't work, to go to the hospital.

"He was never told of the dangers of uranium. When he was in Cove, all those people he worked with passed on. In Oakspring, four of the people he worked with passed away."

PHOTO: MURRAE HAYNES

DAN N. BENALLY, age seventy-seven, of the Red Valley area started working in the uranium-vanadium mines in the Cove area in 1943 for Kerr-McGee. He worked in Mesa mine Nos. 1, 2, 3, 4, 6, and 7. His interview was translated from Navajo by Lorraine Johnson on February 12, 1993, in the Red Valley Chapter House.

"When he was in the mines, the rocks collapsed on him. One of the rocks tore the skin off his side and stomach, too. They had to do a skin graft. He lost part of his eyesight in his left eye."

Benally stopped working in the mines in 1961 because of an accident in which he broke his hips and legs. He worked for the Vanadium Corporation of America in the Oakspring area, in Slickrock and Dove Creek, Colorado, and Monticello, Utah.

"He was never warned how it would affect him in the future. They used to eat underground and drink the water dripping from the walls.

"When they did the blasting, they inhaled the smoke and dust. He fainted twice, and they had to drag him out.

"He worked with thirty-eight people in Cove, and all of them died. One of his relatives passed away recently.

"He was never told or warned [of the dangers]. He was not told to wash [his] hands and [was] told to stay underground." Benally worked shifts from 7 A.M. to 8 P.M. with one hour off for lunch, and earned $1.15 an hour.

He has been X-rayed twice, and a dark spot was found on a lung. "It never goes away. He was told it might be from the uranium, but he was not told if it is cancer. He says he is losing his hair, and he itches a lot."

Benally stopped the interview to reveal a sore on his leg that won't heal. "He was told it might be [from] the uranium. It started two months ago, and it won't go away. It itches a lot." He also has had eye surgery and can't see out of his right eye.

"In the springtime he can still taste the gunpowder."

PHOTO: MURRAE HAYNES

MORRIS GEORGE, age sixty-six, worked in uranium mines for six and a half years beginning in 1953 for Kerr-McGee in the Cove area. The interview was translated from Navajo by Lorraine Johnson on February 12, 1993, in the Red Valley Chapter House.

"He was never told what the things would do to [his] health until recently. If they had told him of the dangers it would cause... he would not have done it. Now he regrets having worked in the mines.

"He's breaking out in a rash that itches. He thinks it's from the uranium. He had an X-ray two years ago, but they never told him the results so he figures he's OK."

George was told to work in Cove. "He was told the uranium was being used to build ammunition for overseas.

"He had no choice but to support his kids back then." He was paid $1.25 an hour. Anglos were the supervisors. "They would only go in the mines to observe."

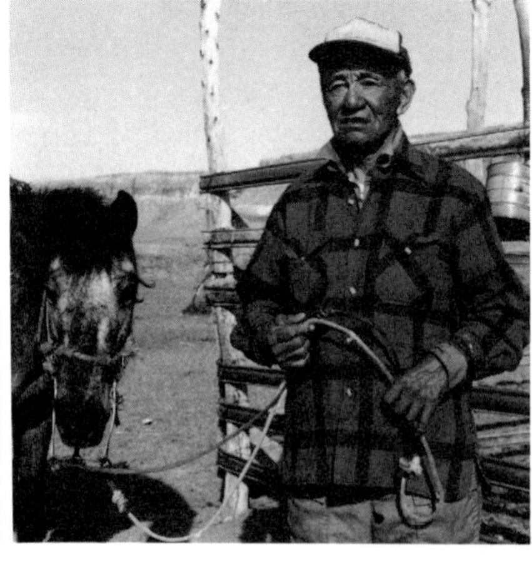

PHOTO: MURRAE HAYNES

LUKE YAZZIE, age eighty-two, led geologists in 1943 to a small rock outcropping near where he was born in an area called Cane Valley, twenty miles east of Monument Valley. What he showed them became a large and very productive mine. It was so productive that a processing mill was built beside his house in Cane Valley. Yazzie has worked in that mine and others all his life and has lived beside the tailings pile since it was created. Until February 1994, he was in excellent health. This interview was conducted in two different places, initially after a Department of Energy public meeting in the Mexican Hat Elementary School on February 18, 1993, and later on Saturday, March 6, at his home in Cane Valley. Luke speaks only Navajo, and his translations were done by Marvin Yellowhair and then later by his daughter, Mary Yazzie, daughter-in-law, Violet, and son, Terry Yazzie.

"He found the uranium ore in the area. It [the area where he found it] used to be [his] playground. He was raised there.

"Before he found out that uranium was hazardous, he benefitted by working in mines as a driller and preparing the blasting caps. But it has been declared dangerous. He worked with uranium twenty-eight years. He has no health problems. His health is monitored by the Public Health Service. So far they have found nothing wrong.

"From his point of view, it is not dangerous, but if it is for other people, then it is a good thing [that the uranium mill tailings are being removed].

"Before the mine, they used the water in the natural springs, and it was not dangerous. The springs were holy. It [the water] should be clean, and we should be able to use it. It is good that [the water] is going to be restored and safe."

Yazzie is concerned that some of the natural wells are drying up.

"He thinks the mining activity in the area affected the natural springs. He'd like to see a well drilled nearby for domestic use. When they drilled the well during the leaching process, that may have dried up the springs. [Water was mixed with acids to wash the crushed ore and leach the uranium out of it]."

The following is Yazzie's account of uranium development next to his home. It begins about 1943.

"There [were] a lot of white people looking for rocks. When he was young, he used to play with it." In the late 1940s and early 1950s the word spread across the reservation that white men were looking for and willing to pay for a certain kind of rock. Samples were to be taken to Goulding's trading post in Monument Valley. Luke took a sample.

"The store owner [Harry Goulding] understood Navajo." Geologists were impressed with the high level of radioactivity in the rocks that Yazzie brought and asked Yazzie to take them to the source. "Then another came from Durango, and with the store owner [Goulding] they crawled into a cave and that is where he had found the rocks. Within a year he [Yazzie] was working in the mine near Goulding's. The ones [the rocks] that he brought were the highest in all the uranium. He started working there the next day.

"He turned twenty-eight when they started working. They taught him how to blast the rocks and make the [cuts]. There was an interpreter there for him. He [Yazzie] was the one who did all the blasting.

"They told him that what he was working on was not that dangerous. It was not true. They had him do the dirty work. He did the job really good and was never hurt. When the job started, many people came here [to Cane Valley]. They were living all around on the rocks. They used up all the trees. Before that there was just the summer camp and the winter camp up in the rocks. There was a store here. Oscar Sloan's wife used to run a store here. There was somewhat over 100 workers.

"They had little trains in the mines, and that is how they got the ore out. He was in there all the time. He never went outside."

Why did the mining company close? "They learned the Navajos were learning about the law, and then they left."

What about the royalties Yazzie expected for his discovery? "He never got them, he said."

And before the mining? "There was only his family [in the valley]. The other families that are here are those who stayed.

"While they were working, Mr. Goulding got rich and moved to Phoenix.

"The workers were told they would get a portion of the profits. But after it was found, they [the white men] didn't tell anyone.

"He [Goulding] built the hotel, the lodge, the hospital, too. When he [Yazzie] went to Goulding, he was told, 'your money is coming.' He never knew what was going on. He [Goulding] used to feed him [Yazzie] and stuff, and act like [he] liked him.

"When the mine stopped, he went to work elsewhere. They [the mining company] never really left anything behind when it was all done.

"The first time they found out it [the uranium] was dangerous was when they started the cleanup about three years ago." (From the late 1960s until 1993, the Yazzie family and neighbors lived next to the 1 million cubic yards of uranium mill tailings that exposed all to radiation.)

"The kids played up there [on the pile] in the summer. We sledded on that during the winter. That['s] the first place I [Violet] went. He [Luke Yazzie] used to get all muddy when he was doing the drilling.

"Nobody said nothing about it from the time they stopped working until the cleanup.

"About twenty [friends, family, and fellow workers] have died, probably more he doesn't know about. A lot of ladies passed away before their husbands. They'd get some kind of infection [cancer?] from inside.

"Many never cleaned their hands, but he [Luke] always cleaned his hands."

Terry Yazzie, Luke's son, who was employed on the crews moving the mill tailings from the valley, enters and begins translating and talking about how the miners spent their pay.

"It went mostly to feeding their kids, and it went to the upkeep of their vehicles. Many lived in shacks and tents all through the valley. A lot of people wasted it [the money]. They bought clothes. Everybody should thank them because they got a lot of money out of it [the work]. In particular, he helped out with the war. He helped out with the country. He wishes he could get a little compensation before he dies. He'd like something to happen so he can see it.

"He wants the sand [mill tailings] taken out so he can do something with the land. Maybe plant corn. We used to play on it [the tailings]. It was good exercise. We'd dig holes in it."

Why has Luke Yazzie escaped any health effects?

"He was in really topnotch shape. He was outside all the time."

Note: In late 1993, Luke Yazzie was diagnosed with restrictive lung disease and spent most of February 1994 in a hospital. By spring he had qualified for compensation and was awaiting payment.

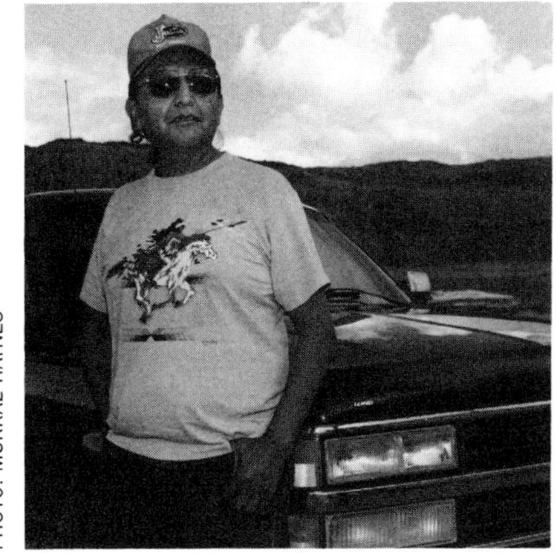

PHOTO: MURRAE HAYNES

BEN STANLEY, about age forty-five, is Luke Yazzie's nephew and also lives adjacent to the Monument Valley mill tailings. He worked in the mill for five years before it closed and has bitter memories about the whole thing. He currently is a watchman for the tailings. This interview was conducted at his home on March 6, 1993.

"In 1960, the mill started here. In 1962, I started working there. I worked there until it closed in 1967. The working conditions were very poor. I started as a laborer, and then I was a crusher for a while. All that dirt came into your body and hair. There were no masks and no protection.

"A rock fell on my toe and crushed my toe. They said they couldn't do anything about it. I had a wife and eleven kids to support. All my toes have healed. The wages were like $1.25 an hour. By the end of eight hours, you were full of dust."

The trucks that hauled the rock from the mine to the mill, a short trip of a mile across a bumpy road of rock and sand, "had no brakes and no starters." The workers had to roll-start them. Ben Stanley worked with Daniel Yazzie, who was Luke's son.

"To the end of the job it was very poor." Not only did his children also play on the tailings pile, but the wastewater from the mill was emptied into a settling pond which the animals used for water. "We butchered those animals and ate them."

Stanley swears he can smell the tailings. "You can smell the tailings and what was mixed with it.

"I have a daughter who lives in North Carolina and gets a rash when she comes back.

"All the houses built in the valley are built with the tailings."

Stanley says that Luke Yazzie's father, an old Navajo named Adakai, warned Luke not to show the geologists where the uranium rocks were. "He said, 'Never take them to the white men. If you do, you'll get nothing out of it.' Luke took [them] to Goulding. He got a cigar for it."

When the mine and mill were in full swing, "they even had a preschool here and three to four houses. The only thing they left behind with us is the trash dump and a lot of old cars half rusted.

"The reason no one ever bothered with us is that we're a small community. Now they [the government] say it is dangerous for the people to work here. What about the people who have lived out here all their lives? They ought to give them all a health exam to see if they're all right."

His brother, Ned Yazzie, now dead, also worked at the mill but was paralyzed when one of the dilapidated trucks he was driving flipped over on him and crushed him.

"We have nowhere else to go, nowhere to turn to.... Everyone who has left has come down with cancer. They have all died."

PHOTO: PETER EICHSTAEDT

IN 1941, BIG JOHN, now age seventy, began working in uranium mines about twenty miles south of the Teec Nos Pos, a small community in the extreme northeast corner of Arizona on the Navajo Reservation. He was interviewed on February 28, 1993, at the Mexican Water Chapter House. His comments were translated by his son, Bill.

"He hauled rock out with a wheelbarrow." There were about fifty people working in the mine. "If they blasted, it was still dusty," when he went into the mine. He was told "none" when he asked if there was any danger to the work. "There was a war going on overseas. They were making some kind of chemical weapons," he was told. He worked a regular shift, from 8 A.M. to noon, with an hour off for lunch, and from 1 P.M. to 5 P.M. They often ate lunch in the mines.

"Nobody told us to wash [our] hands. We never took showers after working."

Big John worked in the mines for about a year, then went to Flagstaff for work, then later worked at a Vanadium Corporation of America mine in Monument Valley beginning about 1948. He worked with Luke Yazzie, who had led geologists to the site. "We never made more than $1.88 an hour. Most of the money was spent on food and transportation."

Big John worked for eighteen years at Monument Valley in both the mine and the Monument Valley mill. His son Bill was born at Monument Valley. Big John did rock blasting as well as hauling the rock out of the mines.

"Nobody told him over the eighteen years that it was dangerous. He is now hard of hearing. Sometimes it is difficult to breathe. His lungs hurt.

"He regrets it [having worked in the mines]. If he knew it [that it was hazardous to his health], he would never have taken the job. He didn't know it would affect his health later on.

"He is worried that he won't get any money. He would be satisfied if he would get a little money from the government."

PHOTO: PETER EICHSTAEDT

BEN JONES, in his late sixties, began working in the uranium mines in 1956 near Naturita, Colorado, and spent eleven years in the mines. He would like to get some compensation from the government for his dangerous work. He was interviewed at the Mexican Water Chapter meeting on February 28, 1993.

He was in the underground, a driller. He said, with the help of a translator, "The dust stayed in the air a long time. You could smell the gunpowder. When you blew your nose, it was yellow dust.

"Sometimes they had vents, but [they were] not strong enough to move the air out." Jones explained that when the generator ran out of gas, "they just left it off.

"Nobody told him it [the work] was dangerous. About six years ago, he first heard about it [the dangers]. His heart has been affected. He had a screening. He has a lump in his heart.

"When he starts walking, he has difficulty breathing. But when he sits down, he is OK.

"He didn't know it was dangerous at the time," he says. "Nobody told him what it [the uranium] was for.

"He wouldn't take another job like that. He has been mostly farming now."

PHOTO: MURRAE HAYNES

WILLIAM LOPEZ worked in the uranium ore processing mill at Tuba City from September 1959 until May 1967, when the operation closed and was eventually dismantled. Lopez is among a large group of former millers seeking inclusion under the Radiation Exposure Compensation Act, from which they have been excluded—preventing them from obtaining $100,000 in compensation payments.

Lopez, about age sixty, was interviewed on February 28, 1993, in the Mexican Water Chapter House and again on May 12, 1993.

"We were milling uranium from the Grand Canyon," he said. The conditions in the mills were dusty and dirty, he said. The ore chunks were crushed to granular size, then soaked in acids and other chemicals to leach the uranium out of the ore. Lopez said he worked in all aspects of the mill.

"It was crushed to a finer form and put into a water solution and with sulfuric acid. That leached the uranium out of the ore. The stripped ore was piled into tailings piles."

Lopez said the processing involved several stages of treatment involving nitric acid solution and ammonia. Once the leachate was collected and the liquids evaporated, "it left yellowcake," Lopez said.

"They had dust scrubbers and vacuums," but these were inadequate, he said.

"We were exposed to all variations of radiation in the mill. Our fingernails were filled with yellow dust. We were breathing yellow dust. Our clothes were covered with the dust. When we swept up, it was a yellow cloud."

Lopez believes that millers were treated unjustly by being excluded from the RECA.

"If the law doesn't include uranium millers, [they] should be included."

Lopez has not suffered any health problems so far, but he knows of many former millers who have and have died.

"I'm fortunate to be as healthy as I am. I know two others who have died and we feel it was from cancer from the mill."

PHOTO: MURRAE HAYNES

GEORGE BROWN, age fifty-five, an employee of the El Paso Natural Gas Company, worked in the Tuba City uranium processing mill from 1959 to 1967. He believes he was exposed to high levels of radiation when the uranium ore was crushed, leached with acids, and then dried into yellowcake. Brown said he repeatedly has asked company officials for his employment and health records but has been told the records are unavailable. His sister, Dorothy Redbird, died in 1981 from cancer of the colon. She worked in the mill, and he believes her death was related to her work at the mill. Brown was interviewed near Window Rock in May 1993.

"She had two kids. I'm trying to fight for her. She's been gone since 1981, but the doctors have to pinpoint the disease she had at the time.

"I started out with the company in 1959 and stayed with them until 1967. I went back and asked my employer for my records. They don't seem to remember my ever working for them. They don't have any of my records. There's quite a few of us [former mill workers] around. I'm still going to try to get some records.

"It's been very hard. The company has changed hands several times. We did a lot of damage to the reservation, and now the government has to come in and clean it up.

"I worked in a real bad place. We shifted around. I started out as a laborer. Back then there [were] hardly any jobs on the reservation. I ended up as a shift foreman."

The mill was divided into certain areas, each a step in the refining process. The crude ore was ground into sand, then soaked with a leaching agent, usually acid; then the uranium was precipitated out and sprayed on hot drums where the liquid would evaporate. The uranium oxide, a fine yellow powder, was then scraped off the drums and funneled into barrels, where it was packed by using a vibrator that helped settle the powder.

"We tried to put 1,000 pounds in a barrel. There were vibrators to pack it and that created a lot of dust.

"They would test our urine. But they never gave us the results. They would vacuum the air and test it, but we were never told [the results].

"We worked with acids, ammonia.... This was all dusty. There were fumes in there. It really stunk. There was no ventilation. This was a danger, but no one ever told us at the time. After I made shift foreman...for about six years I worked with the really hot stuff.

"I used to come out with yellowcake all over my clothes and shoes. My kids played with the shoes and shoelaces. My daughters got cancer of the lymph nodes. My sister used to do all the laundry. She came down with cancer and after treatment had a relapse. I think the treatment killed her.

"We'd like to get something going for my sister. They said you needed medical records. I went to the hospital, and they said I need a court order. There is so much red tape, I wonder if it is worth it. I get so mad sometimes.

"The miners were not exposed to stuff as bad as we were exposed to. It was 90 to 95 percent pure uranium. It was all in the air when we were trying to pack it."

Brown said he has high blood pressure and gets a physical every year. His wife, however, died of cancer. He is convinced that the work around uranium can cause a wide variety of cancers that are not necessarily just respiratory. "What they are looking at is just the respiratory stuff.

"Back in the 1950s they never told us anything. They were doing a lot of atomic testing, and all the dust was falling in the Tuba City area. But they're just looking at the miners. We brought a lot of that stuff home. We never knew it would affect us on down the road. We really don't know what the effect is of this radiation. A few years later it might affect the kids, too. Most of the Anglos I worked with died of things like brain tumors."

PHOTO: MURRAE HAYNES

Shone Holiday, about age seventy-five, lives on a rounded bluff in the heart of Monument Valley, Utah. He lives in a traditional Navajo adobe hogan, has simple corrals for his horses and cows, and from the bluff can watch his flock of sheep. In the afternoon and evening he sits under a structure that provides shade and enjoys some of the most spectacular views in the world. He began working in the uranium mines in the area in the 1940s, and continued to work in uranium and other kinds of mines in the Southwest for nearly thirty years. Standing on the edge of an open-pit mine, he granted this interview in May 1993.

"I've been taking it easy for twenty years now. My lungs are not good. A lot of guys got killed down there. A lot of white guys."

Holiday remembers a cousin of his, Paul Begay, who died in the nearby Starlight uranium mine when it collapsed, crushing him instantly. Holiday worked in the Moonlight mine, one of the largest in the area, which in the summer of 1993 was being filled in after being abandoned for some forty years.

"A lot of guys worked here. About 200 men. We made about a dollar an hour. Sometimes we'd drink water in there. It was cool in the summer time. It was the same in Colorado." He pointed to a large, rusted scooper called a "slusher" that was left at the bottom of the pit. "It was a basket we used to pull it [the ore] out. At that time was World War II. They used [uranium] for the bullets for overseas. I was twenty to twenty-five years old. I was in the army, too. When I came back, I worked in the mines. They drilled with water. Sometimes it was dry, and there was dust. Sometimes the [air] compressors didn't work, and [the mine] would fill up right quick with dust."

Holiday has annual checkups at the Indian Health Service hospital but suffers from shortness of breath. "I can't walk a long way. I used to ride wild horses, but I'm not strong enough now."

PHOTO: MURRAE HAYNES

CECIL PARRISH, about age seventy, lives near Shone Holiday and began working in the mines in 1952. He lives in a traditional Navajo hogan near several modern homes occupied by his children and grandchildren. This interview was conducted near his house in May 1993 with the translation assistance of his son, Wayne.

"We have a lot of mines in the area. He was doing all the blasting. He worked in the area mines until 1960 then went to Colorado. Every now and then he would move around when a mine was played out.

"He didn't have any protection [in] the mines. In the 1940s, they had World War II. It was like an emergency for these miners. There was no warning. They wanted the miners to work twenty to twenty-four hours a day. They got extra pay for drilling and blasting. Back then there was no warning as to the health effects on their life. One of his cousins called Nelson Parrish had a heart problem and passed away. There was another guy who had a heart problem. Now and then he goes to the hospital for health checkups.

"At the Moonlight mine, they worked with bare hands and shoveled the uranium into a bucket and hauled it up. He didn't have any coveralls. They'd come back with uranium all over them. We had no showers back then. It was hard work. They worked you like a slave.

"At first it [the Moonlight mine] was a small hole. A bunch of low-grade ore is still there. They're saying you have to die first before you get a [compensation] check."

PHOTO: MURRAE HAYNES

DON YELLOWHORSE, age forty-nine, began his mining career at age eighteen working in mines operated by Kerr-McGee in the early 1960s in his hometown of Cove, Arizona, not far from where he lives today. This interview was conducted at the Red Valley Chapter House, Red Valley, Arizona, April 10, 1994.

"Almost eight years I worked in the mines in Arizona, New Mexico, Utah," he said. His main job was "taking the rocks out," a job termed "mucker" in the industry.

"When we first came, it [the work] was twenty-four hours a day. It was also dusty and no air. We just started to work. Sometimes a day shift, sometimes at midnight."

Safety equipment and warnings of potential dangers from the collapse of the soft rock or radiation and radon were nonexistent, he explained. "They didn't tell us anything about it," he said of the uranium exposure. "They forced them [the miners] to get the high-grade ore."

Yellowhorse said that mechanical ventilation was rare in the mines, and with the constant drilling and blasting it was often hard to breathe. "In some mines there [were] close to forty people. There was a lot of smoke. We had a hard time getting into the fresh air.

"They had plastic flexible pipes to blow the smoke," he said, but the systems were inadequate. "They [the miners] had a hard time working because of the smoke. You could only see about three feet."

The water-cooled drills were supplied by water tanks, and the leaking hoses provided a source of drinking water, he said. "The water was leaking from the drilling into a pop can wedged under the water [hose]. I drank that a lot of times. It was fresh and cold."

Yellowhorse said his father, John, also was a uranium miner and died of lung cancer at the University of New Mexico Hospital in Albuquerque. "It was about four different things from the uranium," Yellowhorse said of the doctors' explanation of his father's cause of death. "He [his father] worked in the mines for about twenty years."

PHOTO: MURRAE HAYNES

NELSON BENALLY, age sixty-one, was interviewed at the Red Valley Chapter House, Red Valley, Arizona, April 10, 1994. His comments were translated by Helena Benally.

Benally said he remembers "a lot of smoke" and drinking any water available in the mines, water for the drills and water that dripped from the mine walls. He worked in mines on the Colorado Plateau and on the Navajo Reservation. He recalled a mining accident.

"One day the [rock] came down and hit his back. He got hurt. He had to stop working. There was no safety up there." Later he returned to mining and worked periodically over the years in uranium mines around the Red Valley area.

Today, he has difficulty breathing and believes the cause is his work in the mines. He also suffers leg and back pains from mining accidents.

Dangers to health posed by uranium were never explained. "There was nothing," he said. And regarding mine ventilation, he said, "There was some," but it was inadequate.

When asked if he suspected the uranium was a cause of health problems, he said, "He thinks it is...."

PHOTO: MURRAE HAYNES

RUSSELL JACKSON, age seventy-one, worked in mines in the Oakspring area, which is in the northern part of Red Valley, and in Colorado. Sometimes he worked for a Navajo mine operator named Paul Shorty and later for the Vanadium Corporation of America. The interview took place at the Red Valley Chapter House, Red Valley, Arizona, April 10, 1994; his comments were translated by Helena Benally.

The mining crews ranged from twenty to forty men. "There was a lot of dust. Sometimes they worked twenty-four hours 'round the clock."

Dangers in the mines were not a concern, he said. They worked "just to make money."

No one explained why the uranium was being mined. "He didn't know what it was for."

Today, he suffers chest pains whenever he exerts himself, a condition he has had for five or six years.

A year ago he filed a claim for compensation with the government but has yet to receive any money.

PHOTO: MURRAE HAYNES

HARRY JOHNSON, age fifty-seven, worked for ten years in uranium mines—two years in Colorado and the rest in the Cove area. Interviewed at the Red Valley Chapter House, Red Valley, Arizona, April 10, 1994, his comments were translated by Helena Benally.

"He used to take out the rock" and was also a driller. He remembers the smoky, dusty conditions.

"They [would] blast it, [and] he would go back in fifteen minutes later. It was smoky."

Johnson does not remember hearing any warning about the dangers of uranium mining. "They just went in and worked. I guess nobody asked. They [mining companies] never told."

Ventilation pipes were in the mines but apparently were inadequate. "They had it, but it didn't help."

Johnson said he is sorry he worked in the mines. "He regrets it. He shouldn't have worked there."

When asked if he felt he was misled or lied to, he said, "Yes." When asked if he believes he should have been warned, he said, "Yes."

PHOTO: MURRAE HAYNES

HENRY MARTIN, age fifty-eight, worked for about five years in uranium mines in the Cove area and later for two years in Colorado. He was interviewed at the Red Valley Chapter House, Red Valley, Arizona, April 10, 1994; his comments were translated by Helena Benally.

His main job was as a mucker—breaking up the rocks after drilling and blasting, then shoveling the rocks into carts that were hauled out of the mine. He was injured badly in a mine collapse and suffered head, shoulder, and hand injuries. His lasting impression of the mines is of a lot of smoke and dust.

He said he was never warned of the potential health threat that exposure in the uranium mines presented. "He was never told there was danger."

"He gets sick now," it was explained, and he has filed a claim for compensation. However, the Department of Justice has required that he provide check stubs as proof of his employment. Since it was about twenty-five years ago, he has no such records. He is stymied in his request for compensation.

PHOTO: MURRAE HAYNES

FANNIE YAZZIE, approximately seventy years of age, was interviewed at the Red Valley Chapter House, Red Valley, Arizona, April 10, 1994; her comments were translated by Helena Benally.

Yazzie's husband died many years ago, Benally explained, after twenty-five years as a uranium miner. She gave birth to fifteen children, but two died at very young ages of unexplained causes. They lived in a remote area of the reservation and lacked transportation to take their sick children to hospitals.

Yazzie suffers from chronic health problems such as chest pain and weak and aching joints, all of which she attributes to the exposure to uranium that she and her children received from her husband. For her condition, "they just give her a pill," she said of the doctors at the Indian Health Service hospital in Shiprock.

Yazzie is worried about three of her children, who also suffer from unexplained and complicated health problems. She blames the uranium.

Yazzie explained that the family had no running water, and she occasionally washed her husband's work clothes by hand along with the family clothes. "She used to hand wash them."

Frequently her husband did not change clothes when he came home from work, she said. "When he came home, he used to sleep in the clothes in the house."

"She thinks it [uranium] is part of the health problem with the kids."

PHOTO: MURRAE HAYNES

Lawrence Marshall, age forty, was interviewed at the Red Valley Chapter House, Red Valley, Arizona, April 10, 1994.

Marshall was celebrating his birthday on the day of the interview. Like his father, he worked in the mines—beginning in 1972 at the age of eighteen. He worked in a uranium mine in Slick Rock, Colorado, for four years, stopping in 1976.

Because the federal government imposed radon standards beginning in 1970, miners who worked in uranium mines after that date are ineligible for government compensation.

Marshall believes this is unfair because he claims most mining companies did not improve conditions as required by law.

"It was still the same. Smoky and dusty. They passed a law then. That is why I am ineligible," he said.

Marshall said he usually worked a 7 A.M. to 4 P.M. shift and did not leave the mine during the shift. "We stayed down there till four o'clock. We had lunch down there." Although the mining company provided ventilation, it sometimes was inoperable. "There was a vent bag. When it rained, it knocked out the electricity." In such instances, the ventilation system and the lights failed. "We tried to get to the top for fresh air," he recalled.

The conditions in the mines were hot and dusty, he said. "We used to drink the drilling water." Although the mines had ventilation, these mechanical air systems often did not reach to the furthest sections of the mines, where the men worked on extending the ore veins. "In the back rooms, they didn't have ventilation." Often crews working in different shafts did not communicate. As drilling and blasting continued, it often caught nearby crews by surprise, he said. "Sometimes we were still in there when they were blasting," he explained.

Marshall has lung problems, just as his father does. "I have restrictive lung disease. He does, too," he said of his father, Raymond. Marshall also suffers from the memory of a cave-in during which he almost lost his life. "It has bothered me for years," he said. "I'm under mental health. I almost died. It was terrible."

PHOTO: MURRAE HAYNES

AMOS JOHN, age seventy-two, began his mining career in 1948. He continued to work for many years in mines in Oakspring and in Colorado. He worked largely as a general laborer. Interviewed in the Red Valley Chapter House, Red Valley, Arizona, April 10, 1994, his comments were translated by Helena Benally.

Today, he suffers pains in his face and eyes. He got into an accident, he explained. His most vivid memories of mining are dust and smoke. His legs also bother him, and he attributes his failing health to uranium's impact. "When he sits down, it bothers him. It's best when he walks around. He has a hard time breathing. He gets very dry."

He was never warned of the dangers in the mines. "He never asked and was never told." There are many men from Red Valley he worked with in the uranium mines who have died, he said. "The ones he worked closely with are all dead."

Why did he continue? "It was just a job," he said.

PHOTO: MURRAE HAYNES

RAYMOND MARSHALL, age sixty-seven, is the father of Lawrence Marshall. He worked for more than twenty years in uranium mines on the Navajo Reservation, in Colorado, and in Utah, including the VCA No. 2. in Cane Valley. He applied for and has received the $100,000 compensation approved by Congress. He was interviewed in the Red Valley Chapter House, Red Valley, Arizona, April 10, 1994.

"I operated machinery, like a jackhammer. Yes, it was dusty. There was nothing at that time," when it came to safety requirements, he said. "They knew when the mine inspectors were coming. They'd clean everything... put the [blasting] caps away."

Not all mines lacked ventilation, he said. "Some of them were better. They had ventilation." But, he added, "in the old mines it was pretty dusty." He continued to mine until 1975. After years in the uranium mines, he went to work for the coal-burning, electric power plant near Shiprock, New Mexico.

"Right now I have problems with pneumonia. It just won't go away."

Although Marshall qualified for and received his compensation money, he feels he should have been warned of the dangers that he faced in the uranium mines. When asked if he would go back, he said, "No, it's too dangerous. The companies should have told me that."

He received his $100,000 payment in November 1993, he said, and "it is still in the bank."

Appendix I

SIXTY-SIXTH CONGRESS.

SESS. I. CHAPTER 4.1919

Sec. 26. That the Secretary of the Interior be, and hereby is, authorized and empowered, under general regulations to be fixed by him and under such terms and conditions as he may prescribe, not inconsistent with the terms of this section, to lease to citizens of the United States or to any association of such persons or to any corporation organized under the laws of the United States or of any State or Territory thereof, any part of the unallotted lands within any Indian reservation within the States of Arizona, California, Idaho, Montana, Nevada, New Mexico, Oregon, Washington, or Wyoming, heretofore withdrawn from entry under the mining laws for the purpose of mining for deposits of gold, silver, copper, and other valuable metalliferous minerals, which leases shall be irrevocable, except as herein provided, but which may be declared null and void upon breach of any of their terms.

That after the passage and approval of this section, unallotted lands, or such portion thereof as the Secretary of the Interior shall determine, within Indian reservations heretofore withheld from disposition under the mining laws may be declared by the Secretary of the Interior to be subject to exploration for the discovery of deposits of gold, silver, copper, and other valuable metalliferous minerals by citizens of the United States, and after such declaration mining claims may be located by such citizens in the same manner as mining claims are located under the mining laws of the United States: *Provided*, That the locators of all such mining claims, or their heirs, successors, or assigns, shall have a preference right to apply to the Secretary of the Interior for a lease, under the terms and conditions of this section, within one year after the date of the location of any mining claim, and any such locator who shall fail to apply for a lease within one year from the date of location shall forfeit all rights to such mining claim: *Provided further*, That duplicate copies of the location notice shall be filed within sixty days with the superintendent in charge of the reservation on which the mining claim is located, and that application for a lease under this section may be filed with such superintendent for transmission through official channels to the Secretary of the Interior: *And provided further*, That lands containing springs, water holes, or other bodies of water needed or used by the Indians for watering live stock, irrigation, or water-power purposes shall not be designated by the Secretary of the Interior as subject to entry under this section.

That leases under this section shall be for a period of twenty years, with the preferential right in the lessee to renew the same for successive periods of ten years upon such reasonable terms and conditions as may be prescribed by the Secretary of the Interior, unless otherwise provided by law at the time of the expiration of such periods: *Provided*, That the lessee, may in the discretion of the Secretary of the Interior, be permitted at any time to make written relinquishment of all rights under such a lease and upon acceptance thereof be thereby relieved of all future obligations under said lease.

That in addition to areas of mineral land to be included in leases under this section the Secretary of the Interior, in his discretion, may grant to the lessee the right to use, during the life of the lease, subject to the payment of an annual rental of not less than $1 per acre, a tract of unoccupied land, not exceeding forty acres in area, for camp sites, milling, smelting, and refining works, and for other purposes connected with and necessary to the proper development and use of the deposits covered by the lease.

That the Secretary of the Interior, in his discretion, in making any lease under this section, may reserve to the United States the right to lease for a term not exceeding that of the mineral lease, the surface of the lands embraced within such lease under existing law or laws hereafter enacted, in so far as said surface is not necessary for use of the lessee in extracting and removing the deposits therein: *Provided*, That the said Secretary, during the life of the lease, is hereby authorized to issue such permits for easements herein provided to be reserved.

That any successor in interest or assignee of any lease granted under this section, whether by voluntary transfer, judicial sale, foreclosure sale, or otherwise, shall be subject to all the conditions of the lease under which such rights are held and also subject to all the provisions and conditions of this section to the same extent as though such successor or assign were the original lessee hereunder.

That any lease granted under this section may be forfeited and canceled by appropriate proceedings in the United States district court for the district in which said property or some part thereof is situated whenever the lessee, after reasonable notice in writing, as prescribed in the lease, shall fail to comply with the terms of this section or with such conditions not inconsistent herewith as may be specifically recited in the lease.

That for the privilege of mining or extracting the mineral deposits in the ground covered by the lease the lessee shall pay to the United States, for the benefit of the Indians, a royalty which shall not be less than 5 per centum of the net value of the output of the minerals at the mine, due and payable at the end of each month succeeding that of the extraction of the minerals from the mine, and an annual rental, payable at the date of such lease and annually thereafter on the area covered by such lease, at the rate of not less than 25 cents per acre for the first calendar year thereafter; not less than 50 cents per acre for the second, third, fourth, and fifth years, respectively; and not less than $1 per acre for each and every year thereafter during the continuance of the lease, except that such rental for any year shall be credited against the royalties as they accrue for that year.

That in addition to the payment of the royalties and rentals as herein provided the lessee shall expend annually not less than $100 in development work for each mining claim located or leased in the same manner as an annual expenditure for labor or improvements is required to be made under the mining laws of the United States: *Provided*, That the lessee shall also agree to pay all damages occasioned by reason of his mining operations to the land or allotment of any Indian or to the crops or improvements thereon: *And provided further*, That no timber shall be cut upon the reservation by the lessee except for mining purposes and then only after first obtaining a permit from the superintendent of the reservation and upon payment of the fair value thereof.

That the Secretary of the Interior is hereby authorized to examine the books and accounts of lessees, and to acquire them to submit statements, representations, or reports, including information as to cost of mining, all of which statements, representations, or reports so required shall be upon oath, unless otherwise specified, and in such form and upon such blanks as the Secretary of the Interior may require; and any person making any false statement, representation, or report under oath shall be subject to punishment as for perjury.

That all moneys received from royalties and rentals under the provisions of this section shall be deposited in the Treasury of the United States to the credit of the Indians belonging and having tribal rights on the reservation where the leased land is located, which moneys shall be at all times subject to appropriation by Congress for their benefit, unless otherwise provided by treaty or agreement ratified by Congress: *Provided*, That such moneys shall be subject to the laws authorizing the pro rata distribution of Indian tribal funds.

That the Secretary of the Interior is hereby authorized to perform any and all acts and to make such rules and regulations not inconsistent with this section as may be necessary and proper for the protection of the interests of the Indians and for the purpose of carrying the provisions of this section into full force and effect: *Provided*, That nothing in this section shall be construed or held to affect the right of the States or other local authority to exercise any rights which they may have to levy and collect taxes upon improvements, output of mines, or other rights, property, or assets of any lessee.

That mining locations, under the terms of this section, may be made on unallotted lands within Indian reservations by Indians who have heretofore or may hereafter be declared by the Secretary of the Interior to be competent to manage their own affairs; and the said Secretary is hereby authorized and empowered to lease such lands to such Indians in accordance with the provisions of this section: *Provided*, That the Secretary of the Interior be, and he is hereby, authorized to permit other Indians to make locations and obtain leases under the provisions of this section, under such rules and regulations as he may prescribe in regard to the working, developing, disposition, and selling of the products, and the disposition of the proceeds thereof of any such mine by such Indians.

Sec. 27. That hereafter no public lands of the United States shall be withdrawn by Executive Order, proclamation, or otherwise, for or as an Indian reservation except by act of Congress.

Sec. 28. That during this Congress those members of the Committee on Indian Affairs of the House of Representatives, not less than five in number, who are Members of the Sixty-sixth Congress, are authorized to conduct hearings and investigate the conduct of the Indian Service, at Washington, District of Columbia, and elsewhere, and the sum of $15,000, or so much thereof as may be necessary, to be immediately available, is hereby appropriated for expenses incident thereto. The said committee is hereby authorized and empowered to examine into the conduct and management of the Bureau of Indian Affairs and all its branches and agencies, their organization and administration, to examine all books, documents, and papers in the said Bureau of Indian Affairs, its branches or agencies, relating to the administration of the business of said bureau, and shall have and is hereby granted authority to subpoena witnesses, compel their attendance, administer oaths, and to demand any and all books, documents, and papers of whatever nature relating to the affairs of Indians as conducted by said bureau, its branches, and agencies. Said committee is hereby authorized to employ such clerical and other assistance, including stenographers, as said committee may deem necessary in the proper prosecution of its work: *Provided*, That stenographers so employed shall not receive for their services exceeding $1 per printed page.

Approved, June 30, 1919.

Appendix II

AN INTERIM REPORT
of a HEALTH STUDY OF THE URANIUM MINES AND MILLS
by the FEDERAL SECURITY AGENCY
Public Health Service, Division of Occupational Health
and the COLORADO STATE DEPARTMENT OF PUBLIC HEALTH

May 1952

Report prepared by:
Senior Sanitary Engineer Duncan A. Holaday
Surgeon (R) Wilfred D. David
Senior Sanitary Engineer Henry N. Doyle

FOREWORD

The Division of Occupational Health, U.S. Public Health Service, in conjunction with the Colorado Department of Public Health, has been conducting a health study in the uranium mines and mills since July 1950. Although the study is far from complete, sufficient information has been derived to conclude that certain acute conditions are present in the industry which, if not rectified, may seriously affect the health of the worker. We therefore believe it essential that an interim report be issued to the operating companies, describing the progress of the study and presenting methods for correcting any potentially harmful conditions.

It has been gratifying to note that the operating companies, as a result of oral recommendations by our technical personnel, have already started to correct some of these conditions. These diligent efforts must be continued, however, to free the mines and mills of all major health hazards.

I wish to take this opportunity to express my appreciation to the operating companies for their assistance and cooperation in the conduct of this study and to solicit their continued aid in future studies that must be made in order to successfully culminate this project.

Seward E. Miller, *Medical Director*
Chief, Division of Occupational Health
U.S. Public Health Service

TABLE OF CONTENTS

I
INTRODUCTION

Between 1881 and 1887, ores, later found to be radioactive, were discovered on the Colorado Plateau in the Rocky Mountain area of the United States. These were carnotite ores which contained vanadium and uranium, with a small quantity of radium. Several mines were developed during this period, but production was small as there was little or no demand for these materials. A small amount of ore was exported to Europe, and it is reported that the Curies used American carnotite ores in their early experiments.

In 1912, the U.S. Bureau of Mines surveyed the carnotite deposits of the Colorado Plateau and reported them as a practical source of uranium ore. The Bureau of Mines also made a study of the methods for recovering radium, vanadium, and uranium from these ores and operated a pilot plant to demonstrate methods of production. From 1916 to 1923, the domestic production of radium from carnotite ores was significant. In 1923, however, when radium became available from high grade Belgian Congo ores, the low grade American ores were no longer competitive.[1,2,3]

The development of atomic energy focused attention and interest on the carnotite deposits of Colorado and Utah. During the life of the Manhattan Project, many of the old tailings dumps were reworked for their uranium content. Beginning about 1946, the discovery of many new ore bodies in the Colorado Plateau caused the industry to mushroom in this area. At the present time, there are an undetermined number of mines mining uranium ores of various types on the Colorado Plateau. These ores are the carnotites, which have been recognized for many years, roscoelite, autunite, uranite, and torbernite. The State of Colorado presently has the greatest production of uranium ores, with sizeable quantities also mined in Utah, New Mexico, and Arizona. New discoveries, as yet undeveloped, have been reported in Montana and Wyoming. There are eight mills engaged in the processing of the uranium ores, which produce as their final products uranium oxide and vanadic acid. The uranium oxide is shipped to other establishments controlled by the Atomic Energy Commission for refining and subsequent separation of the isotopes.

In August 1949, when the industry had reached a sizeable production rate in the State of Colorado, Dr. Roy L. Cleere, Executive Secretary, Colorado Department of Health, appointed an advisory board to advise him and the State Division of Industrial Hygiene on the procedures to be used in conducting studies and surveys in this industry. Among the conclusions drawn at the first meeting of this group were that little or nothing was known of the health hazards of the uranium producing industry, and that a medical reconnaissance survey should be made by a physician from the Division of Industrial Hygiene* of the U.S. Public Health Service.

Such a reconnaissance survey was made shortly after this meeting. On August 25, 1949, Dr. Cleere called a second meeting of the Advisory Committee, at which was also represented the management of the larger companies mining and producing uranium in the State of Colorado. This group concluded that, in view of the dearth of available information on the health hazards associated with this industry, the Division of Industrial Hygiene of the U.S. Public Health Service should be requested to conduct a study of the uranium mines and mills. Accordingly, on August 30, 1949, a formal request for such a study was made to the Surgeon General by the Colorado Department of Health, the Colorado Bureau of Mines, the Colorado Industrial Commission, and several companies engaged in uranium mining and production. Shortly thereafter, the Chief of the Division of Industrial Hygiene, U.S. Public Health Service, replied to these agencies, expressing the desire and willingness of the Division to conduct a study of this type.

While these negotiations were under way, the Industrial Hygiene Field Station** of the Public Health Service conducted a very limited study of the mining problems in the mines located on the Navajo Indian Reservation. The preliminary information obtained in this brief study indicated that the miners were exposed to external radiation, radon gas, and a high silica dust containing an undetermined amount of radioactivity. In addition, it appeared that the uranium and vanadium contained in the dust were of toxicologic importance and also that the matter of internal radiation due to the inhalation of radioactive dust should be considered.

Shortly thereafter, a preliminary survey was made of a mill producing uranium oxide and vanadic acid. The findings revealed that the mill workers were exposed to uranium- and vanadium-containing dust and to fume and dust of the isolated uranium and vanadium. It was determined that radon was of little significance in the mills because of the large area available for dilution of the gas; that external radiation was not an apparent problem; but that it would be necessary to consider the internal radiation factor, since dust control in the mills was not too effective.

With this background of information, it was concluded that the health study should consist of two essential and correlated phases: a medical investigation and an environmental investigation. Approximately 700 workers were given medical examinations in the first phase of the study, extending from July to October 1950. During the 1951 season, approximately 460 men were examined, of whom 260 were miners and 200 were mill workers. During the two seasons, physical examinations have been performed on 1,117 persons. This

---

* Now known as the Division of Occupational Health

** Now known as the Field Station, Division of Occupational Health

is the maximum number of men that could be readily examined at this time, for the remaining persons are employed at locations so remote that it would be impracticable for a medical team to get to them. However, it is believed that this coverage represents a large enough segment of the working population to permit the drawing of valid conclusions. Since the mines and mills are frequently located at extremely remote locations, it has been necessary to confine the study to the summer and early fall months in order to avoid adverse weather conditions.

The medical study consisted of a detailed physical examination, a chest X-ray, and clinical studies of the blood and urine. The environmental phase included the determination of the nature and concentrations of the materials to which the workers were exposed. Through a correlation of these data, it was hoped that information could be obtained on the health hazards associated with the industry so that necessary control procedures could be instituted.

Environmental studies have been made in six mills and approximately fifty mines. These studies consisted of an evaluation of the dust problem, especially with relation to the silica content of the dust; a determination of the nature and extent of exposure to toxic materials such as uranium, vanadium, and other mineral constituents of the ore; and an evaluation of the radon concentrations in the mines.

Acknowledgments

In order to conduct this study, the Division of Occupational Health, U.S. Public Health Service, has cooperated very closely with the Atomic Energy Commission, especially the Division of Biology and Medicine and the Health and Safety Division of the New York Operations Office. The various State health departments concerned, i.e., Colorado, Utah, New Mexico, Arizona, have been advised of the progress of this study and in many cases have given us valuable cooperation. Assistance has been provided by the U.S. Bureau of Mines, and certain analytical work has been performed by the Los Alamos Scientific Laboratory, the Navy Radiological Defense Laboratory, and the U.S. Bureau of Standards.

Particular acknowledgment is made of the assistance which the Colorado Department of Public Health has given in the conduct of this study. The contribution of the National Cancer Institute through specialized consultation, as well as grants to the Colorado Department of Public Health, is also gratefully acknowledged.

II
EXPERIENCE IN EUROPEAN MINES

In the mines of the Erz Mountains, on both the Bohemian and Saxon sides, it has been known for centuries that the miners die in the prime of life with symptoms of damaged lungs and rapidly progressing ill health. These conditions, especially well known at the Schneeberg mines in Saxony, have been mentioned or described by many writers, including Agricola in 1500, Matthesius in 1559, and Pansa in 1814. It was not until 1879, however, that certain European investigators, through clinical and anatomical research, proved the affliction to be a malignant tumor of the lungs.

On the Bohemian side of the Erz Mountains in Czechoslovakia, about 30 kilometers south and east of Schneeberg, is Jachymov. This small mining town of about 8,000 inhabitants has been world famous as a source of radium from the beginning of the present century, but the history of the mines dates from the 1500's. Early in 1516, rich veins of silver were found in the Jachymov area. Exploitation was then begun but owing to the wasteful methods of mining at that time, the veins were almost exhausted by the end of the 16th century. Later, these mines were exploited for cobalt, nickel, bismuth, and arsenic. In the second half of the 19th century the exploitation began of uranium ore, found chiefly in the form of pitchblende. After Madame Curie's discovery of radium, the pitchblende in the Jachymov mines was processed for radium.

In Jachymov, as in Schneeberg, there has been a high mortality among the miners from pulmonary diseases, notably lung cancer. The similarity between the diseases developed by the miners in these two mining areas was not definitely established until about 1926. Environmental studies have shown that both mines have rather high concentrations of radon because of the presence of radioactive ores.

The causative agent for the lung cancers in these two mining areas has been controversial for a number of years. The etiological agent has been variously attributed to be cobalt, nickel, or arsenic. Others have expressed the view that radon and its decay products are responsible for the cancer of the lungs. It is probable, however, that no single agent can be blamed for this high incidence but a number of combined effects of environmental factors which are difficult to evaluate.

Information available in the medical literature of this country indicates there was an attack rate of about 1 percent per year of lung carcinoma among the miners working in these mines. It is also reported that 50 to 70 percent of all the deaths of the workers in these mines were due to a primary cancer of the respiratory system. This information was the only material available which indicated the health hazards associated with uranium mining.[4,5,6] It must be pointed out, however, that this disease usually has developed only after an average exposure of seventeen years. Moreover, in any attempt to use these findings as a guide, cognizance must be taken of the fact that, in contrast to European practices, American operations are more intermittent. According to the Atomic Energy Commission, generally only one shift is employed, and the mines are not worked on a round-the-clock basis. Consequently, workers are not exposed immediately following the blasting, when dust and radon concentrations are generally highest.

As chronic irradiation per se represents a health hazard regardless of whether or not it induces cancer, all necessary steps have to be taken to reduce this hazard to a practical minimum, since the radon concentrations found in unventilated American mines are comparable to the concentrations found in the Schneeberg and Jachymov mines.

Sufficient information has been obtained in this study to indicate that radon and its decay products can be reduced to a minimum by the provision of adequate and effective ventilation.

III
RADON, ITS BIOLOGICAL AND PHYSICAL PROPERTIES

Uranium ores contain uranium plus all the other members of the radioactive family, of which uranium is the parent. Included in this list is radium, which is transformed into radon gas.

Radon is the heaviest gas known, being about seven times as dense as air. It is absolutely inert chemically and will react with no other material. As it is a gas, it will diffuse from the rocks or be released by drilling and blasting operations and will become dispersed throughout the atmosphere of the mine. It is radioactive and has a half life of about four days, which means that in this period of time one-half of the radon will transmute into other radioactive elements.

The radiation hazard involved in the mining of uranium ores comes from the radioactive gas, radon, and two of its most important daughters, RaA and RaC'. All of these elements emit alpha particles which are very energetic and will damage body cells with which they interact. As radon is a gas, it is breathed in along with the air in the mine and while in the lungs will continue to decay, emitting alpha particles and producing RaA, RaB, RaC, and RaC'. The daughters of radon will also decay in the lungs, likewise emitting alpha particles besides gamma and beta. Furthermore, some of the radon enters the bloodstream. Potential hazard to the lung tissue arises mainly from the alphas from Rn, RaA, and RaC'.

Under usual mine conditions, large numbers of dust particles and water droplets are present in the atmosphere to which the solid decay products of radon will become attached. This dust will be inhaled and carried into the lungs where a portion of it will be retained and decay as outlined above, thus delivering additional radiation to the lungs. The amount of this dust-borne radioactivity that is present in mine atmospheres will depend on the ventilation, air turbulence, and probably other factors. Studies have shown that where ventilation is provided the ratio of radon decay products to radon may be as low as 2 percent of the theoretical value. The facts that the solid daughters of radon will become attached to dust and be inhaled and that the ratio of radon to its decay products is profoundly affected by ventilation are very important in assessing the potential hazard from these elements.

All of these hazards can be effectively reduced by the proper control measures, which are outlined in Section V.

IV
SUMMARY OF MEDICAL FINDINGS

The medical and clinical laboratory examinations of workers in the uranium mining and milling industry in the Colorado Plateau were begun in the summer of 1950. At the present time, over 1,100 men have undergone physical examinations with emphasis upon occupational history, chest roentgenograms, urine analyses, and blood studies, which include erythrocyte counts, total and differential leukocyte counts, hemoglobin estimations, and hematocrit values. In addition, a selected group of approximately 200 miners and millers was examined for atypical blood forms by using a peroxidase staining technique on blood smears.

At this time, no clear-cut etiologic or pathologic patterns have been uncovered among the workers examined. Since the majority of the workers has been exposed for a period less than three years, this observation is not entirely surprising. It does, however, point to the need for repeating the medical studies at frequent intervals. At the present time, therefore, it is planned to reexamine these workers periodically, using the present medical findings as a base line.

In addition, it is felt that a great deal of valuable information could be obtained by an epidemiologic study of persons who were employed in the uranium mines and mills before 1950. At this time, we have not had opportunity to examine a sufficient number of former uranium miners to determine whether or not there has been an excess of lung cancer among them.

An epidemiologic investigator has been assigned to the project to obtain as much information as possible regarding the health status of those persons who were employed in the industry prior to 1950. He is accomplishing this task by obtaining information from such sources as the following: private physicians, referred persons, death certificates, hospital records, industrial commission records, and plant medical and personnel department records. When a relationship between illness and occupation is suspected, arrangements will be made for detailed clinical studies.

The data of the 1950 and 1951 examinations made in the field have been tabulated.

Some of the results are presented here.

Duration of Exposure

Among the total of 913 white miners and millers studied, the following percentage distribution of exposure-durations was found to exist:

EXPOSURE-DURATION	PERCENT OF TOTAL WORKERS
Less than 6 months	23.3
6 months to 1 year	14.4
1 to 3 years	31.4
3 to 5 years	12.4
5 to 10 years	12.7
10 years plus	5.8

Thus, approximately 30 percent of the workers selected have been exposed for a period of three or more years, but within the next 18 months this percentage will rise to approximately 60 percent, provided the turnover of workers does not reach outstanding proportions.

Medical History

Analysis of the medical histories of these individuals showed a predominance of respiratory infections, including pneumonia and sinus infection. There was a predominance of conjunctivitis at the time the physical examinations were performed.

Several cases of illness which were encountered among mill workers were attributed to long-term exposures to relative low concentrations of vanadium compounds.

Dental Findings

A number of workers in the vanadium processing plant were observed to have a green coating of the tongue and teeth. Workers in the area of the uranium leaching process were observed to have a yellow coating of the tongue and teeth. The incidence of dental caries did not differ from that which is observed in the general area. The edentulous persons studied were slightly younger than those previously studied in other parts of the country.

Urine Analyses

Urine analyses have failed to show any significant clinical findings.

Blood Studies

In 1950, approximately 646 determinations were completed on all of the workers examined. In 1951, all of the workers received the same examination with the exception that erythrocyte counts were done on one out of every four persons. Analysis of the blood findings shows no significant abnormalities as far as the red cell count, leukocyte counts, hemoglobin determination, and hematocrit are concerned.

Lungs

In 1950, 13.8 percent of the white miners and 26.5 percent of the white millers showed more than usual pulmonary fibrosis, as compared to 7.5 percent in a control group. In the same year, 20 percent of the Indian millers and 13.2 percent of the Indian miners showed more than usual pulmonary fibrosis, as against none in the controls. Such a finding would indicate a tendency on the part of these individuals to develop silicosis from their exposure. These figures do not imply that the pulmonary fibrosis is occurring because of contact with uranium. It is much more plausible that past exposures to hard rock mining, as well as possible current exposures, are the real cause of this fibrosis. The statement "more than usual pulmonary fibrosis" does not always mean silicosis, but could be one of the early signs of silicosis. Approximately ten cases of definite silicosis have already been determined among the examined workers in 1951.

V
SUMMARY OF ENVIRONMENTAL FINDINGS

During the 1950 and 1951 seasons, fifty mines and eight sampling and processing mills were thoroughly studied in order to determine the toxic materials present and the degree of the workers' exposures. Obviously, individual plant findings cannot be reported in this communication, but the discussion contained herein is, in general, applicable to all milling or mining properties. The investigators have attempted to apprise each company of findings pertaining to its properties.

Mines

Radon and its decay products.—Analysis of a statistically valid number of samples for radon and its short-lived daughters has indicated that the concentration of these radioactive substances has been too high for safe operation over an extended period, except in those locations where adequate ventilation is provided. The median level of radon concentrations in the mines of the Colorado Plateau is above the median levels reported in the European mines, where a high incidence of illness was found in the workers. It must be emphasized again, however, that strict comparison cannot be made between the American and European situations.

A value of 100 micromicrocuries of RaA and RaC' per liter of air* was derived after due consideration of expert opinion and by interpolation of data from other radioactive doses known to produce biological damage. It is believed that the factor of safety in this value of 100 micromicrocuries, as measured by a method to be presented in Section VI, is sufficiently great to prevent damage to the lung tissue in the normal healthy worker. Experiments conducted by this Division indicate that the mines should have no difficulty in reducing the concentration of RaA and RaC' in the mine environment to this suggested value by accepted mine ventilation methods. This value is being offered as an operating standard until such time as a level for total alpha radiation is established by the National Committee on Radiation Protection.

It must be expressly understood that the value of 100 µµc/l** is being suggested without sufficient knowledge on which to base a scientific value and that the Division of Occupational Health reserves the right to reinterpret this value as scientific information is accumulated and when a value is announced by the National Committee on Radiation Protection. Prior studies of environmental hazards in other industries have not had occasion to develop maximum allowable concentration values for RaA and RaC'. The uranium mining and milling industry is the first one in this country in which large numbers of people have been exposed to this contaminant under such conditions that it could not be readily controlled.

* 3.7×10^{10} atomic disintegrations/second = 1 curie

** Micromicrocuries/liter of air

The Division of Occupational Health and the Atomic Energy Commission are presently engaged in the initial phase of an experimental mine ventilation project. When this project is completed, a report of the findings will be issued to the operating companies. Pending the completion of this investigation, the mining companies are urged to immediately install ventilation systems according to standard practices, incorporating the suggestions contained in Section VI of this report.

Dust control is also of great importance in the suppression of RaA and RaC', as the dust particle is a means of attaching the radioactive nuclei to large particles in the atmosphere. Continued emphasis should be placed on wet drilling and the wetting of muck piles. Experiments are in order to determine the effectiveness of wetting agents in the suppression of dust. However, the operators are warned against the use of mine water for wet drilling purposes. It has been found that some mine waters are very high in dissolved radon and that their use constitutes another source of radon in the working atmosphere.

Silica dust.—Most of the uranium encountered in this study has been found in high free silica ore bodies. The free silica content of rafter dust and ore samples has ranged from 40-70 percent. Consequently, the development of silicosis by the miner is a possibility unless adequate dust control measures are used. Several cases of silicosis were detected in the medical study, but all cases had a previous occupational history in hard rock mining. Dust control in the mines has been fairly good, due to the widespread use of wet drilling and because many of the mines are so-called wet mines. Dust concentrations have ranged from 5-20 mppcf (million particles per cubic foot). It is believed that, in view of our present knowledge regarding the development of pneumoconiosis, the silicosis problem in the mines is not acute. The aforementioned range is within the legal limits of most of the States concerned. However, the importance of reducing dust concentrations to a minimum as an adjunct to controlling airborne radioactive materials is again emphasized. Those companies making their own dust counts may experience some difficulty in counting when standard techniques are used. The settling characteristics of the dust are markedly different from those of ordinary mine dust, possibly because of electrostatic charges on the dust particle. Additional information on this subject may be obtained from the Salt Lake City office by persons interested in dust evaluation.

External gamma radiation.—The matter of external radiation received by the miners has been given serious consideration by the study team. Measurements of gamma radiation have been made, using survey meters for area monitoring and pencil dosimeters for personnel monitoring. The results, however, have been erratic. The meaning of the values obtained is obscure, because a number of problems exists in the mines which interfere with radiation measurement. For example, radioactive dust will deposit on any surface and thus concentrate on survey meters and dosimeters. As a result, the instrument readings are primarily functions of the surface of the instrument and the concentration of radioactive dust and not a function of the gamma field produced by the ore bodies. Radioactive dust is also concentrated on the workers' clothes and thus increases the workers' total apparent radiation dose. The problem of measuring external radiation requires considerable study, but this inquiry has been deferred until other more acute problems have been solved. It is strongly recommended that the workers take daily baths (preferably showers) using soap freely, and that a frequent change of work clothing is desirable to minimize skin contact with the radioactive dust.

Internal radiation and chronic metal poisoning.—The levels of atmospheric uranium and vanadium found in the mine atmosphere do not appear to be sufficiently high to produce chronic uranium or vanadium poisoning. Urine samples were taken from all miners examined. The urinary uranium and vanadium levels were of a low order of magnitude.

Internal radiation resulting from the inhalation and absorption of radioactive compounds has been recognized as a potential health hazard because some of the members of the uranium series are bone seekers. This hazard has not been evaluated in the present situation but an investigation is planned. Dust control and ventilation will reduce any possible hazard connected with internal radiation, as well as hazards from chronic metal poisoning.

Mills

Since the process for the separation and isolation of uranium and vanadium is different in each mill, it is impossible to generalize the findings in this report, as was done with the mines, except in a few operations which are common to all plants. In general, it may be said that there are no health hazards in the mills which cannot be controlled by accepted industrial hygiene methods.

Radon and its short-lived decay products are not usual problems in the milling plant, due to the large volume available for dilution and the open-type construction which is, in general, used throughout the industry.

Dust control at the crushing operation has been found to range from fair to poor. The control of dust at this operation can be accomplished by standard engineering methods, but the installation should be designed by competent and skilled persons. Standards set by the State for silicosis control, using the ore of highest free silica content (70 percent), should be used as a guide in determining the effectiveness of dust control systems. This value (5-20 mppcf) should also control other health hazards associated with the dust. Since the radon has not been confined at this operation, the decay products have not had the opportunity to grow into equilibrium. Until adequate dust control has been established at this operation, the workers should be required to wear approved dust respirators. Daily baths and frequent changes of clothing by the workers in this area are also indicated.

Relatively high concentrations of uranium and vanadium fume were found around the fusion furnaces. In practically all plants the workers in this area were found to be suffering from a chronic irritation of the upper respiratory tract, apparently resulting from exposure to vanadium fume. It was also in this area that the green throats referred to in the medical summary were found. Several cases of a transitory illness were observed by the physicians among workers welding or cutting vanadium-coated pipes and metals.

Since the fusion operation is different in each plant, it is impossible to make specific recommendations applicable to each establishment. In general, it may be said that all fusion furnaces should be constructed so as to prevent fume leakage. Local exhaust ventilation should be provided at the transfer point (from furnace to casting wheel) and at the bagging operation. The workers should be provided with fume respirators for emergency and temporary exposures to vanadium and uranium fumes.

Portable exhaust blowers should be located in the fusion area and used by maintenance workers whenever it is necessary to cut or weld metal coated with uranium or vanadium.

Good personal hygiene should be practiced by the workers in the fusion area. This should include daily baths and freshly-laundered work clothes each day.

This report deals only with certain specific hazards in the mill. Exposures to other health hazards, such as acids, alkalies, certain gases, and other agents, were also noted, however. Each mill should therefore consider the advisability of a general industrial hygiene survey by the State industrial hygiene agency or other competent authority.

VI
DISCUSSION OF CONTROL MEASURES

Ventilation of Uranium Mines

The essential requisites for ventilation of uranium mines are the same as those in any siliceous ore mine. These requisites are: planned methods of mine ventilation to provide each work place with an adequate amount of fresh air; wet drilling where practicable; the rapid removal of contaminated air from the workings; and common sense dust control practices.

For operations in rock, emphasis is usually placed on wet drilling, copious use of water to spray down the surface of the work place and muck pile during loading, and forced ventilation to the work face.

The basic health problem associated with uranium mining is not only to suppress the siliceous dusts but also to lower excessively high concentrations of radon and its short-lived daughters. Dust concentrations in these mines are of even greater physiological significance as compared to those in ordinary hard rock mining, since the inspired dust is contaminated with radioactive materials. From the viewpoint of silicosis production, the dust concentration in most uranium mines is within accepted limits. This fact may be explained by the lack of activity in the mines, as they are usually small and generally only one face is worked. Wet drilling is generally practiced, and the mines are usually abandoned overnight to allow powder smoke and dust to be cleared out.

During the course of this study, the ventilation systems of a number of mines have been investigated and, based upon this experience, certain modifications to standard ventilation practice are offered in this report. It is believed that compliance with standard practice and these modifications will result in the lowering of the concentration of RaA plus RaC' to the recommended level of 100 micromicrocuries per liter in practically all mines.

In uranium mining, it appears more practicable to place the mine under positive pressure rather than under negative pressure. This would have several advantages, namely:

1. Churn drill holes from a point close to the breast of development drifts and stopes are becoming more in evidence in the uranium mines. With the mine under positive pressure, radon would be forced from the working area through churn drill holes without distribution throughout the mines.

2. Positive mine pressure would tend to hold the radon in worked-out areas or force it from the mine through surface openings.

3. Although quite theoretical, it may be that positive mine pressure may tend to hold the radon in the fissures and rock openings of newly developed areas. The radon would tend to diffuse out of these areas when the fan is shut off at night (if such be the case), but would subsequently be expelled from these areas 30 or 40 minutes after the ventilation system is turned on.

With regard to nonused workings, it is more desirable to close them off to reduce air contamination and ventilation requirements.

A number of experiments have been performed in uranium mines in order to show the effectiveness of ventilation, and several examples are presented:

1. This mine had two levels. The first level is reached by incline from the surface, and the two levels are connected by means of a second incline. At the time of the ventilation study of this mine, there was approximately 2,800 feet of 8' x 10' drift or tunnel. A 3,000 cfm*fan

* Cubic feet per minute

forced air through a 10" churn drill hole to the lowest level. This air was exhausted to the upper level by way of the incline, providing an air velocity through the drift and incline of 37 fpm.*

Approximately 7,000 cfm of fresh air moved down the main incline to the upper level, mixed with the 300 cfm rising from the lower level, and exhausted from the upper level through a 4' x 6' ventilation raise in which a 10,000 cfm exhaust fan had been installed.

Samples of radon decay products were taken at the foot of each incline. Initial samples were taken when the fans had been inoperative for a period of 14-30 hours. The fans were then turned on and series samples taken over a period of time to observe the effectiveness of ventilation in reducing the concentration of RaA and RaC'. The table below is a tabulation of these data:

SAMPLE LOCATION	RADON DECAY PRODUCTS	REMARKS
Foot # 1 incline	7860 μ μc/l	Fans off 20 hrs.
Foot # 1 incline	4720 μ μc/l	Fans off 30 hrs.**
Foot # 1 incline	Less than 10	Fans off 4 hrs.
Foot # 2 incline	5370 μ μc/l	Fans off 20 hrs.
Foot # 2 incline	5260 μ μc/l	Fans off 30 hrs.**
Foot # 2 incline	Less than 10	Fans on 4 hrs.

2. Another mine consisted of two separate workings in the same ore body. Both workings are reached by incline—Incline # 1 and Incline # 2. Incline # 1 was provided only natural ventilation, and poor air movement was observed. Incline # 2 was under positive pressure from a 2,000 cfm fan which provided 3 to 6 air changes throughout the workings each hour. The concentration of RaA and RaC' in Incline # 1 was in the magnitude of 10,000 μ μc/l, while the concentration in Incline # 2 was 160.

Other examples of ventilation practices are available, but it appears that these two examples serve to indicate what can be accomplished in the reduction of airborne radioactive materials by supplying even small amounts of air to the working area. Pending the completion of the demonstration mine ventilation project being carried on by the Public Health Service and the Atomic Energy Commission, the following general rules may be applied in establishing ventilation requirements for uranium mines:

1. Although each uranium mine will present a special problem in ventilation because of such factors as the grade of ore, amount of radon carried in by ground water, and exposed ore bodies, it is probable that 2,500 cfm is a minimum amount of ventilation required for a small mine.

2. Each pair of men working in a raise, stope, or dead end drift should be supplied 1,000 cfm or more from a tube outlet located within 30 feet of the breast.

3. A supply of not less than 2,000 cfm of fresh air should be supplied to the breast of an 8' x 10' drift.

4. In drifts of large cross section, the quantity of air supply should be calculated to produce a velocity of air flow of not less than 30 fpm.

5. The ventilation system should be turned on at least 40 minutes before workers enter the mine.

---

* Linear feet per minute

** This decrease was caused by the overnight inflow of cold outside air.

6. Natural ventilation of uranium mines should be fully utilized but cannot be relied upon as a suitable means of removing contaminants from the working area.

Collection and Measurement of Radon Decay Products

The success of any control program for radon and its short-lived daughters will depend upon the ability of each mine to make its own measurements. Frequent measurements should be made in the haulageway and daily measurements at the working face. In view of the fact that many workers may spend most of their working lives in uranium mines, it is so important that the suggested level of 100 $\mu\mu c/l$ be maintained in the working environment that industry should not depend on infrequent visits from State or governmental agencies for control work. These agencies should be depended upon only for consultation visits and emergency work. It is noteworthy, also, that transport workers engaged in conveying both the ore from the mines and the more refined product from the mills are not subject to significant exposures.

With these facts in mind, a field method has been developed for the measurements of RaA and RaC', which is relatively simple and may be used by the mines for control work. This method consists essentially of the following steps:

1. Air is drawn through a one-inch diameter circle of Whatman 41 filter paper contained in a special adapter* at a measured rate (14-23 liters per minute) by a hand-cranked pump** for either a 5- or 10-minute period. In order to simplify calculation, tables have been prepared for these two sampling intervals.

2. The filter paper is then removed from the mine and the alpha activity on the paper is measured by a field instrument (the Juno*** is a satisfactory instrument for this purpose). This instrument needs to be calibrated against a laboratory counter so that the scale may be read in alpha disintegrations per minute (dpm). Companies purchasing the instrument should send it to the Salt Lake City office for calibration. The instrument will be returned with a chart converting scale reading to alpha disintegrations per minute, a radioactive standard to check the calibration of the instrument, and a correction curve. In order to prevent contamination, this instrument should never be taken into the mines.

3. The activity (dpm) obtained by reading the instrument must be converted to time zero, which is obtained from the correction curve (see sample calculation below). This value is substituted in one of the following equations:

$$\mu\mu c/l \text{ of RaA and RaC}' = \frac{0.12\ (\text{dpm})}{V} \text{ for 5 minute sample}$$

$$\mu\mu c/l \text{ of RaA and RaC}' = \frac{0.066\ (\text{dpm})}{V} \text{ for 10 minute sample}$$

Where dpm = alpha disintegrations per minute at time zero
V = sampling rate in liters/minute

---

*Blueprint of adapter available on request to Salt Lake City office.

**The only known source of a satisfactory hand-operated pump is Mr. Lester Roberts, 316 Pennsylvania Street, Denver, Colorado.

***This instrument is available from Technical Associates, 3730 San Fernando Road, Glendale 4, California. Cat. No. SRJ-1.

The following is an example of the method:

A sample was taken from 10:03 - 10:13 at a sampling rate of 18.5 liters per minute. The filter paper was read in a Juno instrument at 10:45, giving a reading of 26 x 100. From the calibration chart, alpha disintegrations per minute = 640,000. The correction factor (from a chart to be supplied with instrument calibration) for 32 minutes (10:45 - 10:13) is 1.6; so that the alpha activity at time zero (10:13) was 640,000 x 1.6 or 1,152,000. Substituting this value in the 10-minute formula above, we obtain $\frac{0.066 \times 1,152,000}{18.5} = 4100$ μμc/l of radon decay products.

In order to assist the mining companies in instituting this method of measurement, technical consultation is available by request to the Salt Lake City office. Each set of equipment for these measurements will cost approximately $400.

Address of Salt Lake City Office

Mail

Field Station
Division of Occupational Health
U.S. Public Health Service
Box 2537, Fort Douglas Station
Salt Lake City, Utah

Telephone: 4-2154

Express Shipments

Field Station
Division of Occupational Health
U.S. Public Health Service
Building 503, University of
Utah Annex
Salt Lake City, Utah

VII
RECOMMENDATIONS

Mines

1. Adequate ventilation should be supplied to mines so that the concentration of RaA plus RaC' will not exceed 100 μμc/l at any work location.

2. Wet drilling, wetting of muck pile, and other dust suppressive measures should be practiced to reduce the atmospheric dust concentration to a minimum value.

3. Mine water should not be used for wet drilling purposes unless the water has been certified by a competent authority to be safe for this purpose.

4. Approved dust respirators should be worn by the mine workers until satisfactory ventilation has been provided in the mine.

5. The workers should practice good personal hygiene, including daily showers and frequent change of work clothes. Eating and storage of food in the mine should not be permitted.

6. Pre-employment and periodic medical examinations should be done on all workers. The medical studies should include detailed occupational and medical histories, thorough physical examinations (including skin examinations), laboratory analyses of blood and urine, and chest X-rays. It would be desirable that the periodic examinations be performed annually on every worker.

Mills

1. Effective dust control systems should be applied to the crushing and screening operations.

2. Dust and fume control systems should be used in the fusion and bagging operations at all points where uranium and vanadium dust or fume are liberated.

3. Local exhaust ventilation should be available whenever objects coated with vanadium or uranium are welded or flame cut.

4. Fume or dust respirators should be available to all workers who are exposed to dust or fume until such time as the exposures are controlled by engineering methods. Respirators should be available at all times for emergency or temporary exposures.

5. Good personal hygiene should be practiced by all workers. Daily baths and daily change of work clothes are recommended for those persons employed in the crushing and screening, and the fusion and bagging operations. Separate lunch rooms away from contaminated areas should be provided and used by the workers. No food should be eaten or stored in the production area.

6. Pre-placement and periodic medical examinations should be performed on all workers. The medical studies should include detailed occupational and medical histories; thorough physical examinations, including dermatological examination; laboratory analyses of blood and urine; and chest X-rays. It would be desirable that the periodic examinations be performed annually on every employee.

REFERENCES

1. Liddell. *Handbook of Nonferrous Metallurgy*, Vol. II, McGraw-Hill, 1945, p. 638.

2. Moore and Kithil. A Preliminary Report on Uranium, Radium and Vanadium. *U.S. Bureau of Mines Bull. 70*, 1913.

3. Parsons, Moore, Lind and Schaefer. Extraction and Recovery of Radium, Uranium, and Vanadium from Carnotite. *U.S. Bureau of Mines Bull. 104*, 1916.

4. Thiele, Rostoski, Saupe and Schmorl. Uber den Schneeberger Lungenkrebs. *Munchen. Med. Wechnschr. 71*:24, 1924.

5. Saupe. Uber die Beziehungen zwischen Lungenkrebs und Staublungenerkrankung. *Zentralbl. f. Inn. Med. 54*, 1933.

6. Pirchan and Sikl. Cancer of the Lung in the Miners of Jachymov (Joachimstal). *Amer. Jour. of Cancer*, Vol. XVI, No. 4, July 1932.

Appendix III

HEARINGS BEFORE THE SUBCOMMITTEE ON RESEARCH, DEVELOPMENT, AND RADIATION OF THE JOINT COMMITTEE ON ATOMIC ENERGY CONGRESS OF THE UNITED STATES NINETIETH CONGRESS FIRST SESSION ON RADIATION EXPOSURE OF URANIUM MINERS, 1967

Willard Wirtz's testimony—end of questioning

STATEMENT OF HON. W. WILLARD WIRTZ,[1] SECRETARY OF LABOR

Secretary WIRTZ. Thank you, Mr. Chairman and members of the committee.

I have prepared and had transmitted to the committee a statement covering this matter and would inquire at this point whether it is your pleasure that I read this statement or summarize it in a shorter form.

Representative PRICE. Would you read the statement, Mr. Secretary? The committee is very much interested in your presentation.

Secretary WIRTZ. This committee in calling this hearing has expressed its concern about "the radiation exposure of workmen engaged in the underground mining of uranium ore," and an apparent "lack of needed consummative action to set down limits on such exposure."

The committee's concern is amply justified, and its setting of these hearings a distinctive public service. For these are the facts:

1. After 17 years of debate and discussion regarding the respective private, State, and Federal responsibilities for conditions in the uranium mines, there are today—or were when these hearings were called—no adequate and effective health and safety standard of inspection procedures for uranium mining.

2. There is unmistakable evidence of a high incidence of lung cancer among uranium miners; 98 have died from it, and another 250 to 1,000—the estimates vary—are already incurably afflicted with it.

3. The best available evidence is that over two-thirds of the ap-

---

[1] Biography of Willard Wirtz, Secretary of Labor: Appointed Secretary of Labor by the late President John F. Kennedy in September 1962. Prior to that, had served as Under Secretary of Labor with Secretary Arthur J. Goldberg. Before coming to the Labor Department, career included law teaching, the practice of law, labor arbitration, and public service. At the time of appointment as Under Secretary of Labor in 1961, was a member of the Chicago law firm of Stevenson, Rifkind & Wirtz and a professor of law at Northwestern University.

proximately 2,500 underground uranium miners are working under conditions which at least triple their prospects of dying from lung cancer if they continue this work and these conditions remain unchanged.

DEPARTMENT OF LABOR—RESPONSIBILITY AND ROLE

I have reviewed personally—since the evidence of a high cancer incidence among uranium miners emerged recently, in various reports and in properly diligent press, and even more intensively since these hearings were set—the record which is involved here. Because it offers clear guidance to increased future administrative effectiveness, I summarize it briefly—particularly insofar as it relates to the responsibility and role of the Department of Labor. Any apparent suggestion of a shifting of responsibility will come from inadvertence, for my own association with this situation goes back to June 1961, almost 6 years ago.

The responsibility of the Department of Labor in this area derives from the authority conferred on the Secretary of Labor under the "child labor" provisions of the Fair Labor Standards Act, and under the "health and safety" provisions of the Public Contracts (Walsh-Healey) Act. The Department's less specific but related responsibilities in connection with the development of workmen's compensation programs lie outside the indicated area of the committee's present inquiry.

The Fair Labor Standards Act of 1938 gives the Secretary of Labor responsibility for prohibiting the employment of youth under 18 in hazardous occupations.

In 1942, Hazardous Occupations Order No. 6 was issued, prohibiting employment of those under 18 in "Occupations involving exposure to radioactive substances and to ionizing radiations." This order applies to "work in workrooms" and "any other work which involves exposure to ionizing radiations in excess of 0.5 rem per year." Order No. 6 has not, though, been interpreted as applying to uranium mining.

In 1951, however, in Hazardous Occupations Order No. 9, an earlier (1940) order prohibiting the employment of those under 18 in coal mines was extended to cover "All occupations [specified exceptions not applicable here] in connection with mining other than coal * * *."

Child labor in uranium mines is therefore clearly prohibited under the Fair Labor Standards Act.

The Secretary was born in De Kalb, Ill., on Mar. 14, 1912. He attended Northern Illinois State Teachers College from 1928 to 1930, the University of California in Berkeley (1930–31), and Beloit College in Beloit, Wis. (1931–33), from which he received an A.B. degree. He graduated from Harvard Law School in 1937, and taught law at the University of Iowa and at Northwestern University.

From 1942 to 1946, served in Washington: first as assistant general counsel of the Board of Economic Welfare, then as general counsel and public member of the War Labor Board, and finally as Chairman of the National Wage Stabilization Board.

In addition to Department of Labor responsibilities, chairs a number of governmental committees, including the President's Committee on Manpower; the President's Advisory Committee on Labor-Management Policy (alternating yearly with the Secretary of Commerce); the Missile Sites Labor Commission; and the Interdepartmental Committee on the Status of Women. Is also a member of the Cabinet Committee on Economic Growth, the Federal Development Planning Committee for Appalachia and the Federal Interagency Committee on Education, the President's Council on Aging, the President's Council on Physical Fitness, and the Advisory Council of the President's Committee on Employment of the Handicapped.

Is a director of the American Red Cross and a fellow of the American Academy of Arts and Sciences. Has been awarded honorary degrees from the following institutions: Beloit College, the University of Michigan, the University of Rhode Island, Amherst College, Monmouth College, the University of Iowa, the University of Wisconsin, Northwestern University, Yeshiva University, Ripon College, Roosevelt University, Syracuse University, and Indiana Central College.

Is the author of numerous articles on a wide range of labor topics. His book, "Labor and the Public Interest," was published by Harper & Row in 1964.

There is no evidence of any violation of this prohibition. A review, however, of the extent of inspection activities under these hazardous occupations orders reveals such a limited amount of inspection that an immediate overhaul is being undertaken.

The Public Contracts (Walsh-Healey) Act of 1936 requires that all contracts made by Federal procurement agencies—involving more than $10,000—must include "representations and stipulations" by the contractor that, among other things, "no part of such contract will be performed * * * under working conditions which are * * * hazardous or dangerous to the health and safety of employees engaged in the performance of said contract." If there is a violation of this commitment, the contract may be canceled and the contractor barred from further Government contracts for 3 years.

The administration of the act is in the Secretary of Labor. The key administrative functions, so far as the health and safety provisions are concerned, are (i) the establishing of health and safety standards, and (ii) the inspection of Government contractor premises to assure compliance with these standards.

Since so much of the uranium ore mined in this country is used by mills which have contracts with the Atomic Energy Commission, the Public Contracts Act authority has clear applicability to the uranium mining situation. This has not been questioned, except with respect to certain details regarding the coverage of "independent" mines—those not owned or operated by the milling companies. The AEC contracts with the mills contain broadly phrased "health and safety" stipulations in accordance with the Public Contracts Act requirements.

Yet despite this recognized coverage—and in contrast to the exercise of standard setting and inspection functions by the Department of Labor in almost all other areas of production and commerce—including the processing of uranium ore by AEC contractors—no safety and health standards covering uranium mining have ever been set by the Secretary of Labor—until last week—and no safety and health inspections of uranium mines have ever been made by Department inspectors.

It is a meaningful fact in this record that over the years there was no significant complaint—indeed none at all that I know of—about this inaction by the Department of Labor. The general and prevailing view was that other agencies, Federal and State, could more appropriately exercise here both the standard setting and the inspection functions. There is no excuse, however, for the acceptance by the Secretary of Labor of that view or for the resulting fact that the Public Contracts Act has not been enforced in any case involving the mining of uranium.

The record reflects continuing attention by a variety of State and Federal agencies—including the Department of Labor—to both the standards and the inspection problems in connection with uranium mining. It is a record, nevertheless, of literally hundreds of efforts, studies, meetings, conferences, and telephone calls—each of them leading only to another—most of them containing a sufficient reason for not doing anything then—but adding up over a period of years to totally unjustifiable "lack of needed consummative action."

There were usually, but not always, understandable reasons for the differences of views that developed from time to time about who was responsible for doing what, and authorized to do it—differences—

Between State and Federal agencies about who should inspect the mines;

Between State health and mining authorities;

Between mining associations or companies and State and Federal agencies;

Between the Department of Labor and the Atomic Energy Commission about the scope of applicability of the Public Contracts Act;

Between the Department of Labor and the Department of the Interior about whether uranium mines should be inspected by the agents of one or the other—or whether this should be left to the States; and

Between the Labor Standards Bureau and the Wage, Hour, and Public Contracts Division—both in the Department of Labor—as to where this safety responsibility rested.

Most significantly, no agreement could be reached on a Federal standard covering the permissible degree of radiation exposure in the mines. Various State agencies put standards into effect. There has been substantial opinion in the Federal agencies that these State standards were too loose for human safety, but a variety of opinion as to what would be a right standard.

The testimony of other witnesses at these hearings has and will set this record out in detail which makes my repetition of it unnecessary.

As Secretary of Labor, I issued radiation safety and health standards for Federal supply contracts in February 1964. But these covered only—in general—occupations in the uranium mills and processing operations. They did not cover the mining operations.

The development of radiation exposure standards for the mines came to be more and more recognized as something within the de facto responsibility of the Federal Radiation Council. The Council, as you know, is made up of six members. The Secretary of Labor is one. The efforts of that Council has been reported to you.

Last Thursday—May 4—the Radiation Council convened for what was expected to be the formulation of radiation safety and health standards for uranium mines. The vote within the Council was divided, and no action was taken.

On May 5—the following day—I issued proposed standards under the authority of the Public Contracts Act.

DEPARTMENT OF LABOR—INTERPRETATION OF 0.3 WL STANDARD

The issuance of these standards will be considered wrong by some—on the ground that they are too stringent—precipitous by others.

Postponing this action, at least until after the view of this committee could be obtained in connection with this hearing, had a good deal to commend it. But another delay, even of a few days, is longer, not shorter, after so many years. There was also the belief that the committee's purpose will be served by permitting its consideration not only of why action hasn't been taken, but also of whether the action taken is right. There was most of all the conviction that this situation demands a catalyst to precipitate the results of the countless consultations which have given differences of opinion more than their legitimate influence.

Provision has been made for the receipt of comments on these proposed standards during the next 30 days. Any comments received will be carefully considered, and will be reflected in the standard if this appears appropriate.

The standard issued on May 5 is a 0.3 working level standard.

Well known as this terminology is to the committee, there is perhaps reason to attempt to put this in more generally understandable terms—even at some sacrifice of scientific and medical exactitude.

The problem is to determine how much of the radioactive gas, or dust, which there is in the air in a uranium mine can be breathed in by an individual without creating a special danger of lung cancer. This "gas," or "dust," is, more technically, made up of radon and radon daughters, which, as they decay, emit the alpha radiation which is the cause of the cancer.

There is no disagreement or question but that the inhalation of excessive amounts or concentrations of this radioactive dust do cause lung cancer. Neither is there any question but that some amount of this gas can be inhaled, even indefinitely, without creating an increased lung cancer risk.

Methods have been developed for determining accurately the amount of this radioactive gas in a mine at a particular time. It is measured in terms of the amount of the gas—according to its radioactive effect—which is found in a liter of air in the mine.

If the concentration is equal to a fixed level which has been selected for analytical and discussional purpose—picocuries—this is said to be a 1 working level—or 1 WL—situation. A 3 WL situation would be one in which there is three times that much radioactive gas present; and a 0.3 WL situation is one in which there is about a third—three-tenths—as much radioactive gas as in the 1 WL situation.

Another measurement—working level month—or WLM—is used to cover the time factor in the exposure conditions. Thus, exposure for 170 hours—the average number of working hours in a month—at an average 0.3 WL results in a 0.3 WLM. This would become, on an annual basis 3.6 WLM (12 times the 0.3 WLM).

The basis for adopting the 0.3 WL standard is that the available evidence reflects a widely held view that this represents the maximum level of radioactive material to which a person can be exposed without creating some appreciable increased hazard of lung cancer above what we are all exposed to. Put differently, this evidence is that a man working in a uranium mine where the radioactivity is at or below 0.3 WL is no more likely to get lung cancer than someone else is; if there is more radioactivity than that present, he is more likely to get cancer.[2]

Appendix IV

RADIATION EXPOSURE OF URANIUM MINERS

Report on the Public Health Service Epidemiological Study of Lung Cancer among Uranium Miners (1967 Update)

I. A DESCRIPTION OF THE ORIGIN, ANALYSIS, AND TREATMENT OF ALL EPIDEMIOLOGICAL DATA

Background

Following the determination in 1949 that high concentrations of radon existed in several U.S. uranium mines, and aware of the previously reported association of excess lung cancer mortality among European miners exposed to radon, the Public Health Service in 1950 initiated a long-term study of health hazards associated with the uranium mining industry in the United States. Cooperation and assistance in the study have been received from the Atomic Energy Commission, the Bureau of Mines, the Los Alamos Scientific Laboratory, the Atomic Energy Project of the University of Rochester, and from agencies of the several States concerned. The data collected in this study, which is still going on, relate to both the medical and the environmental aspects and represent essentially the only source of such information in this country.

Epidemiological aspects

The epidemiological investigation includes a base population of 5370 uranium miners and millers in the Colorado Plateau area who volunteered for at least one physical examination during the period 1950–1960 and who provided social and occupational data in sufficient detail to permit followup studies of their health status. Between 1950 and 1960, the uranium miners and millers were given physical examinations in one or more of six years (1950, 1951, 1953, 1954, 1957, and 1960). In 1960 the decision was made to discontinue the repetitive physical examinations because of the high cost, logistical problems, and low yield of clinical findings related to radiation exposure.

In 1960, the application of the sputum cytology technique to the underground uranium mining population was initiated and samples have been collected since then on an annual basis. Miners with suspicious cell pathology are referred to a medical center for further diagnostic studies.

An annual census of uranium miners permits the identification of all active uranium miners and assists in locating those who have left the industry. The census data also permit the construction of an occupational history for each uranium miner which in turn provides the basis for estimating the exposure of uranium miners with underground experience. Miners active at the time of the census also are identified.

Of the 5370 miners and millers, the 3415 white uranium miners with underground mining experience comprised the cohort which was studied intensively. The study group of 3415 white underground uranium miners has been followed closely through the use of a combination of information sources, including the annual census, mail questionnaires, newspaper accounts, employment records, vital statistics bureaus, credit bureaus, the Veterans Administration, and the records of Social Security claims through the Bureau of Old Age and Survivors Insurance. Through this followup program, contact has been maintained with approximately 95 percent of the study group. This type of followup on the study group will be continued for an indefinite period.

Each reported death in the study group is investigated thoroughly. In cooperation with hospitals in the Colorado Plateau area, an intensive effort is made to secure postmortem studies on decreased miners suspected of having lung cancer to ascertain a correct diagnosis, to study the characteristics of the tumor tissue, and to obtain other relevant clinical and pathological information. Other reported deaths in the base population suspected of being lung cancer also are investigated.

The mortality data for the study group have been analyzed by using a modified life table approach. The analyses are based on causes of death as stated on death certificates included under the heading, Malignant Neoplasms of the Respiratory System (Code numbers 160 through 164 in the 6th Revision of the International List of the Causes of Disease). Rates for respiratory cancer and for other causes of death are calculated for the study group and are compared with such rates for the adult white male population of comparable age living in the States of the Colorado Plateau area—Colorado, Utah, Arizona, and New Mexico.

Environmental aspects

A major component of the study has been the measurement of radon and radon daughter concentrations in the mines. At the initiation of the study in 1950, emphasis was placed on radon only, since all previous work had been based on this element. In 1952, when it was determined that the primary health hazard was due to radon daughters rather than radon, the Public Health Service Occupational Health Program developed a field method for the measurement of radon daughters.[1]

Between 1950 and 1956 practically all mine measurements were made by the Public Health Service. Since then other agencies such as the Colorado Bureau of Mines, the New Mexico Bureau of Mines, the Utah Industrial Commission, the U.S. Bureau of Mines, and some mining companies have been making the measurements and providing the data to the Public Health Service. These measurements have been used to calculate cumulative exposures of uranium miners in the epidemiological study and to advise mining companies on procedures for controlling radon daughters.

Summary of Background

Epidemiological data have been derived since 1950 from field observations and measurements in uranium mines and from an intensive long-term followup of a group of underground uranium miners. Data related to other individuals in the mining industry have been collected also. Age standardized mortality rates for respiratory cancer for predetermined exposure groups have been calculated using accented statistical methods. These rates have been compared to mortality rates of white adult males, with due regard to age, living in the same geographic area.

Certain of the data collected during this study, particularly those related to estimates of exposure of individual miners have a certain degree of imprecision. This is due in part to the necessity of using estimated working levels where no measurements are available and in part to procedural errors involved in any sampling study. Nevertheless, these data represent an important source of information regarding U.S. uranium miners, obtained over the period 1950 to the present. There are no other comparable data available.

Analysis of the data has provided the following:

1. A clear demonstration that white underground uranium miners have a higher than expected incidence rate of respiratory cancer which is associated with cumulative exposure to radiation resulting from the radioactive decay of radon.

2. An evident association of excess mortality from lung cancer with estimated exposures of 800 cumulative working level months and above. A demonstrable relationship can be inferred to exist as low as an exposure range of 100–400 cumulative working level months.

3. Although cigarette smoking also may have an effect on the occurrence of lung cancern in underground miners, the analytical comparisons which are presently possible indicate that the observed excess mortality from lung cancer among the study group of uranium miners cannot be accounted for by cigarette smoking alone.

[1] Reference: PHS Bulletin 494, "Control of Radon and Daughters in Uranium Mines and Calculations on Biologic Effects."

II. A DESCRIPTION OF THE ANALYTICAL PREDICTION MODEL INCLUDING ALL SCIENTIFIC AND JUDGMENTAL FACTORS INVOLVED

A major problem in the understanding of the basic mechanism of the biological effect of radiation and of the relationship of continued exposure to the biological effect lies in the infeasibility of controlled experimental trials of the effects of radiation on humans. Data must be drawn from field observations of humans occupationally exposed in a variety of situations. Because of this one must accept the imprecision inherent in such a study. Under these conditions, mathematical models utilizing the data that are available and incorporating certain assumptions can be developed as tools for predicting what may be expected to occur in the future under given sets of circumstances.

Ackoff[2] has stated clearly the limitations which must be considered in constructing an analytical model:

"All models of problem situations are approximate representations of these situations. They are generally simpler than the situations they represent, which are usually so complex . . . that an 'exact' representation (even if possible) would lead to hopeless mathematical complexity. Therefore, the problem confronting the model builder is to attempt a best or good balance between accurate representation and mathematical manageability."

Uranium mining model

The analytical model related to health hazards of uranium miners was developed in an effort to provide a tool which might assist in doing two things:

1. To estimate the number of lung cancers among underground uranium miners which could be expected as a result of exposure to radiation received prior to 1967.

2. To project the lung cancer mortality which could be expected to result from mine operations at various working levels of exposure during the period 1967–86.

Epidemiological data utilized in developing the model

Since experimental trials with humans exposed to radiation are not feasible, data used in the present model had to be drawn from epidemiological field observations. These were obtained from two major sources:

1. A long-term study of U.S. uranium miners carried on by the Public Health Service in collaboration with other Federal and State agencies.
2. Published studies related to Joachimsthal and Schneeberg mines in Europe, Canadian fluorspar miners, and U.S. hardrock (non-uranium) miners.

In the present model, the key variables were considered to be:

1. *Exposure* (WL): The average environmental concentrations of radon daughters, expressed in working levels (WL), to which underground miners are exposed.
2. *Length of exposure* (M): The duration, expressed in months (M) of the exposure of the working population to the stated radon daughter concentrations (WL).
3. *Employment* (E): The average number of miners employed underground during a given period.

Two of these variables, exposure (WL) and length of exposure (M), provide a measure of the cumulative amount of exposure received by an individual, expressed in working level months (WLM). Thus, a miner exposed to 10 working levels for 120 months would have had a cumulative exposure of 1200 working level months (10 WL×120 M=1200 WLM).

The incorporation of the third variable, employment (E) into the formula produces the following identity:

Average working level (WL) × number of men underground (E) × length of exposure (M) or WLEM. The model then considers the formula for the derivation of the number of lung cancers (C) which may occur within a working population to be a function of the interaction of the three variables comprising the WLEM identity, or C=WLEM.

Using data from the five studies noted above, reported lung cancer deaths and sizes of exposed populations were used to calculate three types of indices—

[2] Ackoff, Russell L., *Scientific Method*, 1962, p. 117.

(1) incidence (number of deaths occurring in a given population per unit of time), (2) mortality ratio (ratio of the number of observed deaths to the number that would be expected to occur in the same population in the absence of the specific contributing factor), and (3) the percent of total deaths caused by lung cancer. The values for each of these indices were matched against estimated cumulative exposures for the various population groups of the five studies. From this there was derived the estimate that on the average one lung cancer death could be expected to occur in a population group when the value of WLEM was 5000.

This value was then used as a basic element of the prediction model. Average working levels of radon daughters in U.S. uranium mines were calculated for each of the years 1937 to 1966 (using estimates for the years prior to 1952 and mine measurements and estimates for the years 1952 to 1966). Also, average employment of underground uranium miners for each of the years 1937–1966 was calculated. From these, values of WLEM were determined for each year and divided by 5,000 to estimate the number of lung cancer deaths that could be expected to occur as a result of exposure received during a given year. Using an average period of 20 years from exposure to development of lung cancer, these predicted deaths were distributed over subsequent years from 1956 on through 1984 by allocating 50 percent of the predicted cases at the 20-year lag point and by distributing the remaining 50 percent equally at each of the quartile ranges (i.e., 10 and 30 years). As deaths occur, the validity of the prediction model can be tested against actual occurrence.

In a similar fashion the model was used to predict the numbers of lung cancer deaths which might be expected to occur among underground uranium miners exposed to varying working levels from 1967 to 1986. For this purpose, estimates of underground uranium miner employment during those years were developed and matched against arbitrarily chosen working levels ranging from 6.0 to 0.3. The assumption was made that no miner had been exposed prior to 1967.

III. AN EXPLANATION OF ASSUMPTIONS MADE IN THE FORMULATION AND USE OF THE MODEL

1. That the different groups of miners in the five groups had lung cancer rates which were dependent primarily on radiation exposure and were not greatly influenced by differences in age, ethnic origin, social or other environmental factors. Studies on the U.S. uranium miners have indicated that in the higher exposure levels, at least, the demonstrated dose-response relationship is not attributable to such factors. In the development of this model it has seemed reasonable to assume that the effect of radiation exposure is so great on lung cancer rates that it overwhelms the effect of all other factors involved in lung cancer rates in the mining groups studies.

2. That the atmospheric concentration of radon daughters, expressed in working levels (WL), is directly related to some definite (though unknown) absorbed dose rate of radiation to the tissues of the lung. Atmospheric humidity, size and quantity of dust particles, breathing and smoking habits of each miner and possibly other factors may influence the conversion of exposure to actual dosage in any given case. Though the conversion from exposure rate to dosage rate may not always be constant, the assumption is made that such variation for individuals will occur on both sides of an average value and that the conversion factor will be approximately the same for different groups of miners.

3. That cumulative radiation exposure, expressed as working level months (WLM), is the most relevant radiation parameter that can be correlated with observed lung cancers in an exposed group. Inherent in it is the belief that the biological effect from alpha radiation is related to total dose (i.e., dose rate x time).

4. That a linear relationship exists between radiation dose to a population and its biological effect upon that population. This concept contends that the effect is proportional to cumulative dose with some effect occurring until zero dose is reached. This concept is controversial but many workers in the radiation field accept a linear-nonthreshold relationship, even though the undetectable effect at low dose levels may produce a "practical threshold."

5. That 5,000 working level employment months (WLEM) will on the average produce one lung cancer death. This value was derived from analysis of data from the five studies referred to previously. Such a value could be obtained through a trial and error method by relating different values for WLEM to the lung cancer deaths observed among U.S. uranium miners until a value

is found which results in predicted cases being comparable to cases observed to date.

6. That the average lag time (initial radiation exposure in mines to development of lung cancer) is approximately 20 years. This appears to be the best estimate that can be made at the present time on the basis of the lag periods observed in the Joachimsthal, Schneeberg, and Canadian fluorspar miners. There is a possibility that the lag period may be shorter with higher exposures and longer with lower exposures. It is necessary to use some average lag period value (with a distribution of cases about the mean) to calculate the probable time of occurrence of radiation induced cancers. This also provides a check of the model predictions against the cases actually observed. It should be pointed out, however, that the lag period assumption does not enter into the derivation of the 5,000 WLEM value nor does it enter into the calculation of the numbers of cases being predicted to result from past exposure.

7. That the estimates of average working levels and population sizes at risk are approximately correct. The values have been developed from information given in the literature, surveys of mines, and employment data and they represent the best estimates that can be made at this time.

There are other assumptions which might be incorporated into the model, although doing so would add considerable complexity to the mathematics of the model. For example:

1. That continued exposure to high radiation levels results in "wasted" radiation. That is, if a man continues to be exposed after cancer has been induced, then such exposure cannot contribute to his cancer and is therefore "wasted" from the standpoint of biological effect.

2. That some of the lung cancers predicted by the model may not actually be seen because some individuals destined to develop lung cancer will die from other causes—the so-called competing causes of death. As the lag period lengthens and as a population ages, this competition will increase.

The proposed model has certain serious limitations which preclude its use at this time to predict mortality from lung cancer resulting from past or future experience. Some of these limitations relate to the inherent imprecision of available data. Others relate to the questionable validity or uncertainty of some of the assumptions which are a part of the model, e.g., the lack of knowledge of the interval from exposure to occurrence of lung cancer, the projection of the linear relationship to the low exposure levels, and the choice of a fixed level of cumulative exposure which is equated to the occurrence of lung cancer.

IV. A DESCRIPTION OF THE METHOD USED IN FORMULATING A COMPARISON GROUP FROM THE NORMAL OR CONTROL POPULATION, INCLUDING TECHNIQUES USED TO MAKE ADJUSTMENTS FOR DIFFERENCES IN SUCH THINGS AS SMOKING HABITS

In epidemiological studies such as this, there is not what could be termed a true control population which can be matched against the study population in all characteristics except the one under test. Ideally, in the present study of an occupationally exposed group, the comparison population would be a similar occupational group (underground miners) but lacking the exposure to radiation. This type of control population has not been available in this study.

In part also, the selection of a comparison population is governed by the availability of vital statistics related to that population in a form which permits comparisons to be made with the study group. Vital statistics are often not published in breakdowns which allow the choice of an ideal or nearly ideal comparison population.

For the present study the comparison population is comprised of white males of comparable age living in four States of the Colorado Plateau area—Colorado, Utah, New Mexico, and Arizona. For analysis of mortality data, the modified life-table method was selected. Comparisons with the white male population of the four States were made by means of expected deaths computed from age-race-cause-specific mortality rates for each calendar year and person years at risk among the miners for the same variables.

With regard to the effect of smoking, it has been difficult to obtain published data related to the smoking patterns of the comparison population. However, smoking histories have been collected on the miners in the study group, making possible an internal comparison. From these histories, cumulative cigarette-consumption values, expressed as cumulative pack months, have been calculated (in a manner similar to that by which working level months exposure to radiation is calculated). With the use of person months specific for age and cumulative cigarette consumption, the mortality rates were adjusted for age and cigarette consumption. The exposure response relationship persisted even after this adjustment for smoking so that this factor could not of itself explain the demonstrated exposure-response relationship. As information becomes available on smoking patterns within the comparison population, additional analyses of the impact that cigarette smoking may have on the occurrence of lung cancer among uranium miners will be possible.

V. AN EXPLANATION OF WHAT THE RECOMMENDED STANDARD IS INTENDED TO ACCOMPLISH IN THE LIGHT OF MORTALITY PREDICTIONS

Mortality predictions in the testimony were based upon the mathematical model. Subsequently, limitations in the construction of the model make mortality predictions derived from it subject to serious question. Therefore, it is our feeling that we are unable to make mortality predictions related to a given working level standard.

At the same time we do believe there has been demonstrated a clear relationship of cumulative exposure to excess lung cancer deaths. This, then, is a function of time and level of exposure. At the recommended standard of 0.3 working level, a miner could work for 30 years with cumulative exposure of 120 working level months. Presently available data can be interpreted that a demonstrable exposure-response may occur some place in the 100–400 cumulative working months range. Only time and the continued followup of the study group will provide the data to determine if that value is lower. The choice at this time, therefore, of a working level standard which affords adequate protection to the miners, in the absence of definitive data must be based on judgmental factors.

Appendix V

RADIATION EXPOSURE COMPENSATION ACT

PUBLIC LAW 101-426, 101st Congress
OCT. 15, 1990

An Act

Oct. 15, 1990
[H.R. 2372]

To provide jurisdiction and procedures for claims for compassionate payments for injuries due to exposure to radiation from nuclear testing.

Be it enacted by the Senate and House of Representatives of the United States of America in Congress assembled,

Radiation Exposure Compensation Act.
State listing.
42 USC 2210 note.

SECTION 1. SHORT TITLE.

This Act may be cited as the "Radiation Exposure Compensation Act"

42 USC 2210 note.

SEC. 2. FINDINGS, PURPOSE, AND APOLOGY.

(a) FINDINGS.—The Congress finds that—

(1) fallout emitted during the Government's above-ground nuclear tests in Nevada exposed individuals who lived in the downwind affected area in Nevada, Utah, and Arizona to radiation that is presumed to have generated an excess of cancers among these individuals;

(2) the health of the individuals who were unwitting participants in these tests was put at risk to serve the national security interests of the United States;

(3) radiation released in underground uranium mines that were providing uranium for the primary use and benefit of the nuclear weapons program of the United States Government exposed miners to large doses of radiation and other airborne hazards in the mine environment that together are presumed to have produced an increased incidence of lung cancer and respiratory diseases among these miners;

(4) the United States should recognize and assume responsibility for the harm done to these individuals; and

(5) the Congress recognizes that the lives and health of uranium miners and of innocent individuals who lived downwind from the Nevada tests were involuntarily subjected to increased risk of injury and disease to serve the national security interests of the United States.

(b) PURPOSE.—It is the purpose of this Act to establish a procedure to make partial restitution to the individuals described in subsection (a) for the burdens they have borne for the Nation as a whole.

(c) APOLOGY.—The Congress apologizes on behalf of the Nation to the individuals described in subsection (a) and their families for the hardships they have endured.

42 USC 2210 note.

SEC. 3. TRUST FUND.

(a) ESTABLISHMENT.—There is established in the Treasury of the United States, a trust fund to be known as the "Radiation Exposure Compensation Trust Fund" (hereinafter in this Act referred to as the "Fund"), which shall be administered by the Secretary of the Treasury.

(b) INVESTMENT OF AMOUNTS IN THE FUND.—Amounts in the Fund shall be invested in accordance with section 9702 of title 31, United

States Code, and any interest on, and proceeds from any such investment shall be credited to and become a part of the Fund.

(c) AVAILABILITY OF THE FUND.—Amounts in the Fund shall be available only for disbursement by the Attorney General under section 6.

(d) TERMINATION.—The Fund shall terminate not later than the earlier of the date on which an amount has been expended from the Fund which is equal to the amount authorized to be appropriated to the Fund by subsection (e), and any income earned on such amount, or 22 years after the date of the enactment of this Act. If all of the amounts in the Fund have not been expended by the end of that 22-year period, investments of amounts in the Fund shall be liquidated and receipts thereof deposited in the Fund and all funds remaining in the Fund shall be deposited in the miscellaneous receipts account in the Treasury.

(e) AUTHORIZATION OF APPROPRIATIONS.—There are authorized to be appropriated to the Fund $100,000,000. Any amounts appropriated pursuant to this section are authorized to remain available until expended.

SEC. 4. CLAIMS RELATING TO OPEN AIR NUCLEAR TESTING. 42 USC 2210 note.

(a)(1) CLAIMS RELATING TO CHILDHOOD LEUKEMIA.—Any individual who was physically present in the affected area for a period of at least 1 year during the period beginning on January 21, 1951, and ending on October 31, 1958, or was physically present in the affected area for the period beginning on June 30, 1962, and ending on July 31, 1962, and who submits written medical documentation that he or she, after such period of physical presence and between 2 and 30 years of first exposure to the fallout, contracted leukemia (other than chronic lymphocytic leukemia), shall receive $50,000 if—

(A) initial exposure occurred prior to age 21,

(B) the claim for such payment is filed with the Attorney General by or on behalf of such individual, and

(C) the Attorney General determines, in accordance with section 6, that the claim meets the requirements of this Act.

(2) CLAIMS RELATING TO SPECIFIED DISEASES.—Any individual who was physically present in the affected area for a period of at least 2 years during the period beginning on January 21, 1951, and ending on October 31, 1958, or was physically present in the affected area for the period beginning on June 30, 1962, and ending on July 31, 1962, and who submits written medical documentation that he or she, after such period of physical presence, contracted a specified disease, shall receive $50,000 if—

(A) the claim for such payment is filed with the Attorney General by or on behalf of such individual, and

(B) the Attorney General determines, in accordance with section 6, that the claim meets the requirements of this Act.

Payments under this section may be made only in accordance with section 6.

(b) DEFINITIONS.—For purposes of this section, the term—

(1) "affected area" means—

(A) in the State of Utah, the counties of Washington, Iron, Kane, Garfield, Sevier, Beaver, Millard, and Piute;

(B) in the State of Nevada, the counties of White Pine, Nye, Lander, Lincoln, Eureka, and that portion of Clark County that consists of townships 13 through 16 at ranges 63 through 71; and

(C) that part of Arizona that is north of the Grand Canyon and west of the Colorado River; and

(2) "specified disease" means leukemia (other than chronic lymphocytic leukemia), provided that initial exposure occurred after the age of 20 and the onset of the disease was between 2 and 30 years of first exposure, and the following diseases, provided onset was at least 5 years after first exposure: multiple myeloma, lymphomas (other than Hodgkin's disease), and primary cancer of: the thyroid (provided initial exposure occurred by the age of 20), female breast (provided initial exposure occurred prior to age 40), esophagus (provided low alcohol consumption and not a heavy smoker), stomach (provided initial exposure occurred before age 30), pharynx (provided not a heavy smoker), small intestine, pancreas (provided not a heavy smoker and low coffee consumption), bile ducts, gall bladder, or liver (except if cirrhosis or hepatitis B is indicated).

State listing.
42 USC 2210 note.

SEC. 5. CLAIMS RELATING TO URANIUM MINING.

(a) ELIGIBILITY OF INDIVIDUALS.—Any individual who was employed in a uranium mine located in Colorado, New Mexico, Arizona, Wyoming, or Utah at any time during the period beginning on January 1, 1947, and ending on December 31, 1971, and who, in the course of such employment—

(1)(A) if a nonsmoker, was exposed to 200 or more working level months of radiation and submits written medical documentation that he or she, after such exposure, developed lung cancer, or

(B) if a smoker, was exposed to 300 or more working level months of radiation and cancer incidence occurred before age 45 or was exposed to 500 or more working level months of radiation, regardless of age of cancer incidence, and submits written medical documentation that he or she, after such exposure, developed lung cancer; or

(2)(A) if a nonsmoker, was exposed to 200 or more working level months of radiation and submits written medical documentation that he or she, after such exposure, developed a nonmalignant respiratory disease, or

(B) if a smoker, was exposed to 300 or more working level months of radiation and the nonmalignant respiratory disease developed before age 45 or was exposed to 500 or more working level months of radiation, regardless of age of disease incidence, and submits written medical documentation that he or she, after such exposure, developed a nonmalignant respiratory disease,

shall receive $100,000, if—

(1) the claim for such payment is filed with the Attorney General by or on behalf of such individual, and

(2) the Attorney General determines, in accordance with section 6, that the claim meets the requirements of this Act.

Payments under this section may be made only in accordance with section 6.

(b) DEFINITIONS.—For purposes of this section—

(1) the term "working level month of radiation" means radiation exposure at the level of one working level every work day for a month, or an equivalent exposure over a greater or lesser amount of time;

(2) the term "working level" means the concentration of the short half-life daughters of radon that will release (1.3×10^5) million electron volts of alpha energy per liter of air;

(3) the term "nonmalignant respiratory disease" means fibrosis of the lung, pulmonary fibrosis, and corpulmonale related to fibrosis of the lung; and if the claimant, whether Indian or non-Indian, worked in an uranium mine located on or within an Indian Reservation, the term shall also include moderate or severe silicosis or pneumoconiosis; and

(4) the term "Indian tribe" means any Indian tribe, band, nation, pueblo, or other organized group or community, that is recognized as eligible for special programs and services provided by the United States to Indian tribes because of their status as Indians.

SEC. 6. DETERMINATION AND PAYMENT OF CLAIMS. 42 USC 2210 note.

(a) ESTABLISHMENT OF FILING PROCEDURES.—The Attorney General shall establish procedures whereby individuals may submit claims for payments under this Act.

(b) DETERMINATION OF CLAIMS.—

(1) IN GENERAL.—The Attorney General shall, in accordance with this subsection, determine whether each claim filed under this Act meets the requirements of this Act.

(2) CONSULTATION.—The Attorney General shall—

(A) in consultation with the Surgeon General, establish guidelines for determining what constitutes written medical documentation that an individual contracted a specified disease under section 4 or other disease specified in section 5; and

(B) in consultation with the Director of the National Institute for Occupational Safety and Health, establish guidelines for determining what constitutes documentation that an individual was exposed to the working level months of radiation under section 5.

The Attorney General may consult with the Surgeon General with respect to making determinations pursuant to the guidelines issued under subparagraph (A), and with the Director of the National Institute for Occupational Safety and Health with respect to making determinations pursuant to the guidelines issued under subparagraph (B).

(c) PAYMENT OF CLAIMS.—

(1) IN GENERAL.—The Attorney General shall pay, from amounts available in the Fund, claims filed under this Act which the Attorney General determines meet the requirements of this Act.

(2) OFFSET FOR CERTAIN PAYMENTS.—A payment to an individual, or to a survivor of that individual, under this section on a claim under section 4 or 5 shall be offset by the amount of any payment made pursuant to a final award or settlement on a claim (other than a claim for worker's compensation), against any person, that is based on injuries incurred by that individual on account of—

(A) exposure to radiation, from open air nuclear testing, in the affected area (as defined in section 4(b)(1)) at any time during any period specified in section 4(a), or

(B) exposure to radiation in a uranium mine at any time during the period described in section 5(a). Uranium. Mine safety and health.

(3) Right of subrogation.—Upon payment of a claim under this section, the United States Government is subrogated for the amount of the payment to a right or claim that the individual to whom the payment was made may have against any person on account of injuries referred to in paragraph (2).

(4) Payments in the case of deceased persons.—

(A) In general.—In the case of an individual who is deceased at the time of payment under this section, such payment may be made only as follows:

(i) If the individual is survived by a spouse who is living at the time of payment, such payment shall be made to such surviving spouse.

(ii) If there is no surviving spouse described in clause (i), such payment shall be made in equal shares to all children of the individual who are living at the time of payment.

(iii) If there is no surviving spouse described in clause (i) and if there are no children described in clause (ii), such payment shall be made in equal shares to the parents of the individual who are living at the time of payment.

(iv) If there is no surviving spouse described in clause (i), and if there are no children described in clause (ii) or parents described in clause (iii), such payment shall be made in equal shares to all grandchildren of the individual who are living at the time of payment.

(v) If there is no surviving spouse described in clause (i), and if there are no children described in clause (ii), parents described in clause (iii), or grandchildren described in clause (iv), then such payment shall be made in equal shares to the grandparents of the individual who are living at the time of payment.

(B) Individuals who are survivors.—If an individual eligible for payment under section 4 or 5 dies before filing a claim under this Act, a survivor of that individual who may receive payment under subparagraph (A) may file a claim for such payment under this Act.

(C) Definitions.—For purposes of this paragraph—

(i) the "spouse" of an individual means a wife or husband of that individual who was married to that individual for at least one year immediately before the death of that individual;

(ii) a "child" includes a recognized natural child, a stepchild who lived with an individual in a regular parent-child relationship, and an adopted child;

(iii) a "parent" includes fathers and mothers through adoption;

(iv) a "grandchild" of an individual is a child of a child of that individual; and

(v) a "grandparent" of an individual is a parent of a parent of that individual.

(d) Action on Claims.—The Attorney General shall complete the determination on each claim filed in accordance with the procedures established under subsection (a) not later than twelve months after the claim is so filed.

(e) Payment in Full Settlement of Claims Against the United States.—The acceptance of payment by an individual under this section shall be in full satisfaction of all claims of or on behalf of that individual against the United States, or against any person with respect to that person's performance of a contract with the United States, that arise out of exposure to radiation, from open air nuclear testing, in the affected area (as defined in section 4(b)(1)) at any time during any period described in section 4(a), or exposure to radiation in a uranium mine at any time during the period described in section 5(a).

(f) ADMINISTRATIVE COSTS NOT PAID FROM THE FUND.—No costs incurred by the Attorney General in carrying out this section shall be paid from the Fund or set off against, or otherwise deducted from, any payment under this section to any individual.

(g) TERMINATION OF DUTIES OF ATTORNEY GENERAL.—The duties of the Attorney General under this section shall cease when the Fund terminates.

(h) CERTIFICATION OF TREATMENT OF PAYMENTS UNDER OTHER LAWS.—Amounts paid to an individual under this section—

(1) shall be treated for purposes of the internal revenue laws of the United States as damages for human suffering; and

(2) shall not be included as income or resources for purposes of determining eligibility to receive benefits described in section 3803(c)(2)(C) of title 31, United States Code, or the amount of such benefits.

(i) USE OF EXISTING RESOURCES.—The Attorney General should use funds and resources available to the Attorney General to carry out his or her functions under this Act.

(j) REGULATORY AUTHORITY.—The Attorney General may issue such regulations as are necessary to carry out this Act.

(k) ISSUANCE OF REGULATIONS, GUIDELINES, AND PROCEDURES.—Regulations, guidelines, and procedures to carry out this Act shall be issued not later than 180 days after the date of the enactment of this Act.

SEC. 7. CLAIMS NOT ASSIGNABLE OR TRANSFERABLE; CHOICE OF REMEDIES. 42 USC 2210 note.

(a) CLAIMS NOT ASSIGNABLE OR TRANSFERABLE.—No claim cognizable under this Act shall be assignable or transferable.

(b) CHOICE OF REMEDIES.—No individual may receive payment under both sections 4 and 5 of this Act.

SEC. 8. LIMITATIONS ON CLAIMS. 42 USC 2210 note.

A claim to which this Act applies shall be barred unless the claim is filed within 20 years after the date of the enactment of this Act.

SEC. 9. ATTORNEY FEES. 42 USC 2210 note.

Notwithstanding any contract, the representative of an individual may not receive, for services rendered in connection with the claim of an individual under this Act, more than 10 per centum of a payment made under this Act on such claim. Any such representative who violates this section shall be fined not more than $5,000.

SEC. 10. CERTAIN CLAIMS NOT AFFECTED BY AWARDS OF DAMAGES. 42 USC 2210 note.

A payment made under this Act shall not be considered as any form of compensation or reimbursement for a loss for purposes of imposing liability on any individual receiving such payment, on the basis of such receipt, to repay any insurance carrier for insurance payments, or to repay any person on account of worker's compensation payments; and a payment under this Act shall not affect any claim against an insurance carrier with respect to insurance or against any person with respect to worker's compensation.

42 USC 2210 note.

SEC. 11. BUDGET ACT.

No authority under this Act to enter into contracts or to make payments shall be effective in any fiscal year except to such extent or in such amounts as are provided in advance in appropriations Acts.

42 USC 2210 note.

SEC. 12. REPORT.

(a) The Secretary of Health and Human Services shall submit a report on the incidence of radiation related moderate or severe silicosis and pneumoconiosis in uranium miners employed in the uranium mines that are defined in section 5 and are located off of Indian reservations.

(b) Such report shall be completed not later than September 30, 1992.

Approved October 15, 1990.

Appendix VI

REPORT ON THE URANIUM MINER'S SCREENING PROJECT
SHIPROCK, NEW MEXICO

By Louise Abel, M.D., Shiprock Indian Health Services (IHS)
June 5, 1993

I would like to present our experience in Shiprock, New Mexico, regarding trying to provide medical testing for the Navajo Uranium Miners. I will summarize the results of our testing so far, then I will present our current difficulties with the regulations.

I am a general internist with the Indian Health Services. I have been working part-time for the past 4½ years in Shiprock. I became the primary physician for many of our Uranium Miners in Shiprock when I assumed responsibility for the pulmonary clinic. After the final regulations were published it became clear that the IHS could not provide most of the testing required for living miners. I approached Dr. Allan Craig, Clinical Director, SRSU, and Dr. Doug Peter, Chief Medical Officer, Navajo Area IHS, about this problem. Most of the Navajo Uranium Miners rely on the IHS for all their medical care. We had never even heard of some of the requirements like "B readings." We identified a miner's hospital in Raton, New Mexico, the Miner's Colfax Medical Center, which could provide the testing we needed. They could travel to Shiprock to provide this service via an outreach van, equipped with a full pulmonary testing lab. Mr. Timothy Benally, Director, Navajo Uranium Mining Screening Project, approached us at this time and we decided to coordinate a screening effort using his extensive registry of Navajo miners. The Miner's Colfax Medical Center was hired through a special contract. A public health nurse was hired for four months to coordinate the testing phase. Dr. Jonathan Samet's department at UNM provided "B readers" to read the x-rays.

From November 16 through February 5, 1993, 549 Navajo miners were tested with an x-ray and basic breathing tests. So far, we have 474 readings back (86% of the total). We found that 14% of the miners have Silicosis. Eight percent (8%) of the miners have lung abnormalities severe enough to meet the RECA criteria, but only 2.3% of the miners meet both criteria as required. We discovered four cases of lung cancer. One miner had a cardiac arrest in our office, but was resuscitated.

We divided the miners into four groups based on the results of the two tests:

GROUP I:	Normal CXR, "Normal" PFTs	373
GROUP II:	Normal CXR, "Abnormal" PFTs	32
GROUP III:	Silicosis, "Normal" PFTs	58
GROUP IV:	Silicosis, "Abnormal" PFTs	11

Group IV is the group of miners that meet the qualifications for compensation. Group III have Silicosis, but need further exercise testing to see if they are disabled enough to meet the RECA criteria. Group II have lung disease that is yet undetermined, and will need further evaluation with pulmonary consultation and High Resolution CT scan. Some of these miners may qualify, too. Group I is a diverse group because the "normal" PFTs are in fact a broad range which includes some true normals and some that are mildly abnormal. I think selected subgroups of this category deserve further testing as above.

So we have just gotten started with identifying which miners need further testing and which tests will be helpful. The IHS has no regular funding available for these tests which are only available through referral medical centers. The first round of testing, the simple non-invasive tests, have only identified a fraction of the miners who are disabled.

Through this initial phase of testing we identified several problems with the regulations. It appeared that the x-rays were missing some miners with Silicosis. It became clear that pulmonary testing of the type we did was not sensitive enough to pick up the disability of the miners. The regulations allow for exercise testing, and this will be needed to accurately assess the men. This is a test that is not available at the Shiprock-IHS Facility and we are currently making arrangements.

I presented these concerns to the Justice Department and they convened a panel to discuss them on May 16, 1993. They agreed to accept a new test, High Resolution CT Scan, because the panel agreed the chest x-rays miss 10-20% of miners with Silicosis. I appreciated the chance to have my concerns heard.

Unfortunately, some of the problems could not be addressed because they reflected inconsistencies in the law itself. One particular requirement which has caused some men to be denied is the requirement *that miners with Silicosis have worked on an Indian Reservation*. Medically, the evidence that Silicosis occurs from uranium mining is very solid. Our Navajo miners worked all over the Southwest. Many of these are having difficulty with compensation because *they cannot prove enough exposure solely on the Reservation*. The original law contains the specific requirement that miners with Silicosis must have worked on the Reservation. *This does not make medical sense*. The sandstone is the same all over the Southwest. Silicosis resulted from the work in any of the uranium mines on or off the Reservation. In the regulations this inconsistency was translated in the harshest way. The regulations require that miners have *all* 200 WLM exposure on the Reservation, which makes the miners meet the most extreme interpretation of this nonsensical paragraph.

Another problem that has plagued us is the regulation requiring medical documents to be certified. The IHS (a Federal entity) has had difficulty meeting this requirement due to lack of staff. This has put the Navajos who rely on the IHS at a *disadvantage*. Removal of this requirement would help us alot and not compromise the accuracy of the documents. The Federal Government should accept its own documents without the certification requirement.

In summary, we have tried to meet the needs of the Navajo miners for special testing. We found from our initial studies that the easy non-evasive tests leave many people out of compensation. The more difficult, expensive tests will be required to adequately test the men. The Justice Department has been responsive to my medical concerns; however, two issues remain: *Certification of records and the "Reservation" requirement for Silicosis*.

Notes

Unless otherwise noted, quotations were made at the places and times indicated in the text.

CHAPTER 1

1. Results of the Uranium Miners Screening Project, Wednesday, February 10, 1993, Office of Navajo Uranium Workers, Indian Health Service, Shiprock, New Mexico.

CHAPTER 2

1. A. Q. Lundquist and J. L. Lake (of United Carbide Nuclear Company), "History and Trends of the Uranium Plant Flowsheet," *Mining Congress Journal* (November 1955):38.

2. Gary Lee Shumway, "A History of the Uranium Industry on the Colorado Plateau" (Ph.D. diss., University of Southern California, 1970), 15–20.

3. Ibid., 30–31.

4. Henry Wade, *The Begay Story* (Atascadero, Calif.: Begay Story, 1978).

5. U.S. Senate Committee on Indian Affairs, *Leasing of Indian Lands*, hearings on Senate Bill 145, 57th Cong., 1st sess., January 16–23, 1902, 4–5.

6. Ibid., 6.

7. Ibid., 23.

8. Ibid., 17.

9. Ibid., 42.

10. Ibid., 119.

11. Shumway, "History of the Uranium Industry," 19.

12. Ibid., 21.

13. Ibid., 67.

14. Ibid., 67–70.

15. H.R. 2480, 66th Cong., 1st sess., Chapter 4, Laws of 1919, 31–32. (This was an act making appropriations for the current and contingent expenses of the Bureau of Indian Affairs, for fulfilling treaty stipulations with various Indian tribes and for other purposes, for the fiscal year ending June 30, 1920.)

16. William L. Chenoweth, *Historical Review of Uranium-Vanadium Production in the Northern and Western Carrizo Mountains, Apache County, Arizona*, open-file report 85–13, Arizona Bureau of Geology and Mineral Technology (Tucson, Arizona, June 1985), 9.

17. William L. Chenoweth, *Historical Review of Uranium-Vanadium Production in the Eastern Carrizo Mountains, San Juan County, New Mexico, and Apache County, Arizona*, open-file report 193, New Mexico Bureau of Mines and Mineral Resources (Socorro, New Mexico, March 1984), 3.

18. Shumway, "History of the Uranium Industry," 77.

CHAPTER 3

1. Gary Lee Shumway, "A History of the Uranium Industry on the Colorado Plateau" (Ph.D. diss., University of Southern California, 1970), 92–96.

2. William L. Chenoweth, *Early Vanadium-Uranium Mining in Monument Valley, Apache and Navajo Counties, Arizona, and San Juan County, Utah*, open-file report 85–15, Arizona Bureau of Geology and Mineral Technology (Tucson, Arizona, 1985), 2.

3. Shumway, "History of the Uranium Industry," 97–108.

4. Chenoweth, *Vanadium-Uranium Mining in Monument Valley*, 2–5.

5. Shumway, "History of the Uranium Industry," 118.

6. Ibid., 126.

7. H. L. Anderson, *Bulletin of Atomic Scientists* (September 1974):42.

8. William L. Laurence, *Men and Atoms: The Discovery, the Uses and the Future of Atomic Energy* (New York: Simon and Shuster, 1959), 55–56.

9. Richard G. Hewlett and Oscar E. Anderson, Jr., *A History of the United States Atomic Energy Commission*, vol. 1, *The New World, 1939–1946*, 26, 65.

10. William L. Chenoweth, *Uranium Procurement and Geologic Investigations of the Manhattan Project in Arizona*, open-file report 88–2, Arizona Bureau of Geology and Mineral Technology (Tucson, Arizona, 1988), 13.

11. D. C. Duncan and W. L. Stokes, *Vanadium Deposits in the Carrizo Mountain District, Navajo Indian Reservation, Northeastern Arizona and Northwestern New Mexico*. Prepared for the U.S. Army Corps of Engineers, Manhattan Engineering District, by the U.S. Geological Survey (Washington, D.C., 1942), 19.

12. Chenoweth, *Uranium Procurement*, 15.

CHAPTER 4

1. William L. Chenoweth, *Early Vanadium-Uranium Mining in Monument Valley, Apache and Navajo Counties, Arizona, and San Juan County, Utah*, open-file report 85–15, Arizona Bureau of Geology and Mineral Technology (Tucson, Arizona, 1985), 10.

2. Neilsen B. O'Rear, *Summary and Chronology of Domestic Uranium Program*, U.S. Atomic Energy Commission (Grand Junction, Colorado, May 1966), 4, 5.

3. Ibid., 8.

4. Ibid., 8, 9.

5. Ibid., 10,11,12.

6. Ibid., 13–17.

7. Navajo Tribal Council Resolution, October 14, 1949.

8. Chenoweth, *Vanadium-Uranium Mining in Monument Valley*, 11.

9. William L. Chenoweth, *Historical Review of Uranium-Vanadium Production in the Eastern Carrizo Mountains, San Juan County, New Mexico, and Apache County, Arizona*, open-file report 193, New Mexico Bureau of Mines and Mineral Resources (Socorro, New Mexico, March 1984), 15–18.

10. Paul E. Melancon, *Geology and Technology of the Grants Uranium Region*, Society of Economic Geologists, Memoir 15, New Mexico Bureau of Mines and Mineral Resources (Socorro, New Mexico, 1963), 3, 4.

11. William L. Chenoweth, unpublished interview, December 1, 1990, Grand Junction, Colorado, Oral History Project, and personal communication, March 1993.

12. Ibid.

13. Wilbert L. Dare, *Underground Mining Methods and Costs of Three Salt Wash Uranium Mines of Climax Uranium Company*, U.S. Bureau of Mines, Information Circular 7908 (Washington, D.C., 1959), 32.

14. Ibid., 32, 33.

15. Wilbert L. Dare, *Uranium Mining in the Lukachukai Mountains, Apache County, Arizona, Kerr-McGee Oil Industries, Inc.*, U.S. Bureau of Mines (Washington, D.C., 1961), 9.

16. Ibid.

CHAPTER 5

1. Anna Rondon, personal communication, November 1992.

2. Henry N. Doyle, memorandum, "Survey of Uranium Mines on Navajo Reservation, November 14–17, 1949, January 11–12, 1950," U.S. Public Health Service [hereafter referred to as U.S. PHS] (Salt Lake City, Utah, 1950), 3.

3. Ibid., 3.

4. Ibid., 5.

5. Ibid., 6.

6. Ibid., 2.

7. Ibid., 5.

8. R. H. Allpert, District Mining Supervisor, letter to H. I. Smith, Chief Mining Branch, U.S. Geological Survey, December 14, 1949.

9. Ethelbert Stewart, "Survey of Industrial Poisoning from Radioactive Substances, *Labor Review*, U.S. Bureau of Labor Statistics, 28, no. 6 (June 1929): 20–61.

10. Ibid., 28.

11. Ibid., 22, 23.

12. Egon Lorenz, "Radioactivity and Lung Cancer: A Critical Review of Lung Cancer in the Mines of Schneeberg and Joachimsthal," *Journal of the National Cancer Institute*, 5, no. 1 (August 1944), 5–7.

13. Dr. Bernie Wolf and Merril Eisenbud, *Medical Survey of Colorado Raw Materials Area*, to P. C. Leahy, Manager, Colorado Area Office, Atomic Energy Commission (July 19, 1948), 1.

14. Ibid.

15. Ibid.

16. Ibid.

17. Ibid., 3.

18. Ibid., 3–5.

19. Ralph V. Batie, letter to Stewart Udall, December 26, 1979, 1, 2.

20. Ibid., 2.

21. Henry N. Doyle, Senior Sanitary Engineer, Chief of the Industrial Hygiene Field Station, Public Health Service, Salt Lake City, Utah; letter to Dr. Wilfred D. David, Division of Industrial Hygiene, U.S. PHS, August 8, 1948, 1.

22. Ibid.

23. Duncan A. Holaday, Wilfred D. David, and Henry N. Doyle, *Interim Report of a Health Study of the Uranium Mines and Mills*, Federal Security Agency, U.S. PHS, Division of Occupational Health, and the Colorado State Department of Public Health (May 1952), 6.

24. Ibid., 7.

25. Ibid., 8.

26. Duncan A. Holaday and Henry N. Doyle, *Progress Report (July 1950–December 1951) on the Health Study in the Uranium Mines and Mills*, U.S. PHS (Salt Lake City, Utah, January 1952), 7.

27. Ibid., 9.

28. Ibid., 12.

29. Duncan A. Holaday deposition in *Barnson et al. v. United States*, February 11, 1981, 66. D. Utah C-81005.

30. Holaday and Doyle, *Progress Report*, 10–12.

31. Holaday, David, and Doyle, *Interim Report*, 8.

32. Roy B. Snapp, Secretary, the Atomic Energy Commission, "Radiation Problem in Uranium Mines of the Colorado Plateau," Internal Communication (April 4, 1952), 5.

33. Wilbert L. Dare, R. A. Lindblum, and J. H. Soule, *Uranium Mining on the Colorado Plateau*, U.S. Bureau of Mines, Information Circular 7726 (Washington, D.C., September 1955), 55.

34. Ibid.

35. Ibid.

36. Duncan Holaday, letter to Stewart Udall, May 21, 1983.

37. Duncan Holaday, letter to Stewart Udall, July 20, 1984.

38. Duncan Holaday, letter to Stewart Udall, January 16, 1984.

39. Duncan Holaday, letter to Stewart Udall, January 29, 1979.

CHAPTER 6

1. Minutes of Uranium Study Advisory Committee Meeting, February 26, 1953, Field Headquarters, Division of Occupational Health (Cincinnati, Ohio), 2, 3.

2. Ibid., 4.

3. Lewis L. Strauss, Chairman, Atomic Energy Commission, letter to W. Sterling Cole, Chairman, Joint Committee on Atomic Energy, U.S. Congress (July 13, 1953), 2.

4. Ibid., 5.

5. Ibid., 5.

6. Duncan Holaday, Senior Sanitary Engineer, Acting Chief, Field Station, U.S. PHS, Salt Lake City, Utah, letter, August 27, 1953, to Allen D. Look, Chief, Arizona Section, Health and Safety Service, U.S. Bureau of Mines, 1.

7. Duncan Holaday, Senior Sanitary Engineer, U.S. PHS, letter, September 21, 1953, to Jim DeMoss, Division Safety Engineer, Kerr-McGee Oil Industries, 1.

8. Duncan Holaday, Senior Sanitary Engineer, U.S. PHS, internal office memorandum (September 30, 1953), 1.

9. Ibid.

10. Minutes of the Uranium Study Advisory Committee Meeting, December 1953, Division of Occupational Health, U.S. PHS (Salt Lake City, Utah), 4.

11. Ibid., 5.

12. Seward E. Miller, Chief, Division of Occupational Health, U.S. PHS, memorandum, February 17, 1954, "Pathological and Chemical Studies of Tissues from Autopsied Miners," 1.

13. Howard E. Ayers, report, *Control of Radon and Its Daughters in Mines by Ventilation*, March 15, 1954, Occupational Health Field Station, U.S. PHS (Salt Lake City, Utah), and the Atomic Energy Commission, document AECU-2858, 4.

14. Victor E. Archer, U.S. PHS (Salt Lake City) memorandum, "Permission to Testify at Colorado Hearing, Regarding Medical History of Tom Van Arsdale," October 10, 1957, to Pope Lawrence, Environmental Field Studies Section, U.S. PHS and National Cancer Institute (Washington, D.C.), 3.

15. Ibid., 4.

16. Victor E. Archer, U.S. PHS, *Preliminary Report on the Health Study of Uranium Miners* (Salt Lake City, Utah, September 3, 1958), 3–5.

17. Ibid., 6.

18. Duncan Holaday, Chief, Occupational Health Field Station, U.S. PHS (Salt Lake City, Utah), draft testimony, February 19, 1959, *Radiation Exposures in Uranium Mines and Mills*, before the Joint Committee on Atomic Energy, U.S. Cong. (Washington, D.C.), 1.

19. M. Allen Pond, Henry Doyle, John Johnson, Frank Costello, and Forrest Western, *Report of the Inter-Agency Committee on Health Hazards in Uranium Mining* (Washington, D.C., November 12, 1959), 4.

20. Ibid.

21. Ibid.

22. Ibid.

23. U.S. Bureau of Mines, *Radiation Protection Activities in the Department of Interior*, report for Dr. Arthur S. Flemming, Federal Radiation Council (Washington, D.C., August 9, 1960), 1.

24. *Governors' Conference on Health Hazards in Uranium Mines: A Summary Report*, U.S. PHS Publication No. 843 (Washington, D.C., 1961).

25. Ibid.

26. Ibid.

27. Ibid.

28. Ibid.

29. Henry Doyle, Assistant Chief, Division of Occupational Health, U.S. PHS, hearing testimony, March 21, 22, 1966, Subcommittee on Labor, Senate Committee on Labor and Public Welfare, 20.

30. J. V. Reistrup, "Hidden Casualties of Atom Age Emerge," *Washington Post*, March 9, 1967.

31. Ibid.

32. Sen. John O. Pastore, Chairman of the Joint Committee on Atomic Energy, press release, April 21, 1967.

33. Hearings, U.S. Congress, Subcommittee on Research, Development, and Radiation, Joint Committee on Atomic Energy, 90th Cong., 1st sess., May 9, 1967, to August 10, 1967, 12.

34. Ibid., 45–47.

35. Ibid., 100.

36. Ibid., 102.

37. Ibid., 106, 107.

38. Ibid., 116.

39. Ibid., 125.

40. Ibid., 126.

41. Ibid., 194, 195.

42. Ibid., 218–22.

43. Ibid., 290.

44. Ibid., 290.

45. Ibid., 351.

46. Ibid., 472.

47. Ibid., 479.

48. Ibid., 483.

49. Ibid., 433.

50. Ibid., 521.

51. Ibid., 555.

52. Ibid., 585.

53. Ibid., 588.

54. Ibid., 728–32.

55. Ibid., 736.

CHAPTER 7

1. Urith Lucas, "Navajos Who Mined Uranium Dying from Lung Cancer, Relatives Seeking Federal Compensation," *Albuquerque Tribune*, August 17, 1973.

2. Ibid.

3. "Aid for Navajo Miners Who Have Lung Cancer," *Navajo Times*, October 11, 1973.

4. "Lung Cancer Hearings for Workers in Uranium Mines Are Scheduled," *Navajo Times*, circa November 1973.

5. Bene Goldenberg, "Navajo Family Subsidies Sought, BIA Investigation Begins," *Albuquerque Tribune*, June 14, 1974.

6. Harry Tome, letter to Rep. Harold Runnels of New Mexico, June 1975.

7. Resolution RRC-48–76, Red Rock Chapter of the Navajo Tribe, "Requesting the United States Congress, the Navajo Tribal Council and Indian Health Service to Take Appropriate Action to Assist Former Navajo Uranium Miners and Their Families."

8. Rep. Manuel Lujan, Jr., letter to Harry Tome, Navajo Tribal Council, March 24, 1977.

9. Sen. Pete Domenici, letter to Harry Tome, Navajo Tribal Council, March 25, 1977.

10. Rep. Harold Runnels, letter to Harry Tome, Navajo Tribal Council, April 1, 1977.

11. Harry Tome and Elwood Tsosie, letter to Stewart Udall, September 29, 1978.

12. Stewart Udall, letter to Harry Tome, December 15, 1978.

13. Dan Smart, "Tragedy at Red Rock," *Oklahoma Observer*, February 25, 1979.

14. Chris Shuey, "The Widows of Red Rock," *Saturday Magazine, Scottsdale Progress*, June 9, 1979.

15. Gordon Eliot White, "Why Utah Fallout Aid Is Slow in Coming," *Deseret News*, July 20, 1979.

16. Molly Ivins, "Uranium Mines Leaving Indians a Legacy of Death," *New York Times*, May 20, 1979.

17. Ibid.

18. Arthur R. Tamplin and John W. Gofman, *The Colorado Plateau: Joachimsthal Revisited? An Analysis of the Lung Cancer Problem in Uranium and Hardrock Miners*, testimony, hearings, Joint Committee on Atomic Energy, 91st Cong., January 28, 1970, 1.

19. Ibid., 19, 20.

20. Reed Madsen, "Utah Reluctant to Pay Mine Widows," *Deseret News*, June 6, 1979.

21. Stewart Udall, letter to Harry Tome, July 3, 1979.

22. Stewart Udall, fact sheet, *Uranium Miners Claims Filed with the Department of Energy*, July 20, 1979.

23. Sen. Pete Domenici, statement at U.S. Senate Special Committee on Aging, Grants, New Mexico, August 30, 1979, 1.

24. Testimony, Agnes Ratliff, ibid., 8.

25. Testimony, Dennis Heppler, ibid., 9.

26. Testimony, Stewart Udall, ibid., 16.

27. Testimony, Stewart Udall, ibid., 17.

28. Testimony, Stewart Udall, ibid., 19.

29. Testimony, Harry Tome, ibid., 31.

30. Testimony, Carl Thomas, ibid., 32.

31. Testimony, Marie Harvey, ibid., 34.

32. Testimony, Jessie Harrison, ibid., 36, 37.

33. Testimony, Harris Charley, ibid., 39.

34. Stewart Udall, letter to Joseph DiStefano, Acting Assistant General Counsel, Department of Energy, "Uranium Miner Claims," August 22, 1979, and July 20, 1979; and Ellen Wheeler, "$44 Million Urged for U-Miners, Kin," *Rocky Mountain News*, September 6, 1979.

35. Harry Tome, letter to George Vlassis, General Counsel, Navajo Tribe, December 11, 1979.

36. Al Senia, "Suit Says Indian Miners Were Exposed to Radiation," *Washington Post*, December 15, 1979.

37. Denise Tessier, "Uranium Mine Gas Causes Lung Cancer, UNM Group Told," *Albuquerque Journal*, March 11, 1980.

38. Kathie Saltzstein, "EPA Head Seeks Help for Red Valley Navajos Affected by Uranium," *Gallup Independent*, April 7, 1980.

39. Stewart Udall, letter to Peter McDonald, Chairman of the Navajo Tribe, May 28, 1980.

40. William P. Mahoney, Jr., letter "Re: Navajo Uranium Miners" to Harry Tome, October 3, 1980.

41. Eldon W. Hansen, Controller, Navajo Nation, letter to James Nahkai, Jr., Chairman of Budget and Finance Committee, re: "Proposed Resolution of the Budget and Finance Committee Recommending the Appropriation of $15,000," April 13, 1981.

42. Stewart Udall, "Newsletter to Navajo Uranium Miners and Widows," June 11, 1981.

43. Harry Tome, testimony for the "Field Hearing of the U.S. Senate Committee on Labor and Human Resources, Senate Bill 1483 entitled "Radiation Exposure Compensation Act of 1981," Salt Lake City, Utah, April 14, 1982.

44. Stewart Udall, "Newsletter to Navajo Uranium Miners and Widows," April 12, 1982.

45. Gay Jervey, "Stewart Udall's Newest Frontier," *American Lawyer* (January/February 1991): 78.

46. Ibid.

47. Ibid.

48. Ibid., 79, and personal interview, September 4, 1993.

49. Leonard Haskie, Interim Chairman, Navajo Tribal Council, at hearing, Subcommittee on Mineral Resources Development and Production, of the Committee on Energy and Natural Resources, U.S. Senate, 101st Cong., 2nd sess., *Impacts of Past Uranium Mining Practices* (Shiprock, New Mexico, March 13, 1990), 11.

50. Testimony, Dr. Richard W. Hornung, National Institute for Occupational Safety and Health, ibid., 23.

51. Testimony, Dr. Susan Dawson, Utah State University, ibid., 149, 150.

52. Testimony, Phillip Harrison, Jr., Uranium Radiation Victims' Committee, ibid., 163.

53. Testimony, Earl Mettler, Esq., ibid., 171.

54. Public Law 101–426, October 15, 1990, "Radiation Exposure Compensation Act," section 2, paragraph (c).

CHAPTER 8

1. U.S. Department of Energy, *UMTRA Shiprock Completion Overview* (Shiprock, New Mexico, October 3, 1986).

2. U.S. Department of Energy, *Uranium Mill Tailings Remediation Action Project* (Shiprock, New Mexico, Groundwater Project, February 1993), 2.

3. U.S. Department of Energy, *Shiprock Completion Overview*, 2.

4. Harold Tso, Director, Environmental Protection Commission, Navajo Nation, letter to Sen. Pete Domenici, hearing before the Special Committee on Aging, U.S. Senate, 96th Cong., 1st sess. (Grants, New Mexico, August 30, 1979), 87.

5. Ford, Bacon and Davis Utah, *Engineering Assessment of Inactive Uranium Mill Tailings, Monument Valley Site, Monument Valley, Arizona*, for U.S. Energy Research and Development Administration (Grand Junction, Colorado, March 31, 1977), 1–6.

6. Joseph M. Hans, Jr., and Richard L. Douglas, *Radiation Survey of Dwellings in Cane Valley, Arizona, and Utah, for Use of Uranium Mill Tailings*, for the Environmental Protection Agency, Office of Radiation Program (Las Vegas, Nevada, August 1975), 2, 3.

7. Ford, Bacon, Davis Utah, 1–12.

8. Ford, Bacon, Davis Utah, 1–13, 14.

9. Ford, Bacon, Davis Utah, 1–8.

10. U.S. Department of Energy, *Uranium Mill Tailings Remedial Action Project* (Tuba City, Arizona, Groundwater Project, March 1993).

11. James H. Olsen, Jr., *Jack Pile Reclamation Project—History and Progress Update*, Tailings Waste and AML Reclamation, Billings Symposium, 1993, section VII, paragraph (e).

CHAPTER 9

1. Macario Juarez, Jr., "Radiation Compensation, U.S. Pays First Navajo Claims," *Gallup Independent*, May 28, 1992.

2. Stewart Udall, letter to Stuart Gerson, Assistant Attorney General, and Helene Goldberg, Director, Torts Branch, U.S. Department of Justice, December 17, 1992.

3. Ibid., 4.

4. Stuart M. Gerson, Assistant Attorney General, letter to Stewart Udall, January 14, 1993, 1, 2.

5. Ibid., 5.

6. Ibid., 6.

7. Personal communication, June 1993.

8. Testimony, Maude Todacheene, hearing of the Committee on Labor and Human Resources, U.S. Senate, 103rd Cong., 1st sess., *Examining the Implementation of the Radiation Exposure Compensation Act of 1990* (Shiprock, New Mexico, June 5, 1993), 47.

9. Testimony, Jess White, ibid., 39.

10. Frank Krider, Assistant Director, Torts Branch, U.S. Department of Justice, letter to Stewart Udall, April 7, 1993.

11. Victor E. Archer, letter to Stewart Udall, regarding work history for Keeda Hosteen Nez Begay, May 7, 1993.

12. Frank Krider, "Decision" re: Claim Number 201–16–296, April 5, 1993.

13. Stewart Udall, U.S. Senate hearing "Examining the Implementation," June 5, 1993, 50.

14. Ibid., 51.

Bibliography

BOOKS

Hewlett, Richard G., and Oscar E. Anderson, Jr. *A History of the United States Atomic Energy Commission*. Vol. 1, *The New World, 1939–1946*. University Park, Pa.: Pennsylvania State University Press, 1962.

Laurence, William L. *Men and Atoms: The Discovery, the Uses and the Future of Atomic Energy*. New York: Simon and Schuster, 1959.

Trafzer, Clifford E. *The Kit Carson Campaign: The Last Great Navajo War*. Norman, Okla.: University of Oklahoma Press, 1982.

Wade, Henry. *The Begay Story*. Atascadero, Calif.: Begay Story, 1978.

REPORTS

Abel, Louise. *Report on the Uranium Miner's Screening Project, Shiprock, New Mexico*. Indian Health Services, June 5, 1993.

Archer, Victor E. *Preliminary Report on the Health Study of Uranium Miners*. U.S. Public Health Service. Salt Lake City, Utah, September 3, 1958.

Ayers, Howard E. *Control of Radon and Its Daughters in Mines by Ventilation*. U.S. Public Health Service and Atomic Energy Commission. AECU-2858, Salt Lake City, Utah, March 15, 1954.

Chenoweth, William L. *Historical Review of Uranium-Vanadium Production in the Eastern Carrizo Mountains, San Juan County, New Mexico, and Apache County, Arizona*. Open-file report 193, New Mexico Bureau of Mines and Mineral Resources. Socorro, New Mexico, March 1984.

———. *Early Vanadium-Uranium Mining in Monument Valley, Apache and Navajo Counties,*

Arizona, and San Juan County, Utah. Open-file report 85–15, Arizona Bureau of Geology and Mineral Technology. Tucson, Arizona. Undated.

———. *Historical Review of Uranium-Vanadium Production in the Northern and Western Carrizo Mountains, Apache County, Arizona.* Open-file report 85–13, Arizona Bureau of Geology and Mineral Technology. Tucson, Arizona, June 1985.

———. *Uranium Procurement and Geologic Investigations of the Manhattan Project in Arizona.* Open-file report 88–2, Arizona Bureau of Geology and Mineral Technology. Tucson, Arizona, January 1988.

Dare, Wilbert L. *Underground Mining Methods and Costs of Three Salt Wash Uranium Mines of Climax Uranium Company.* Information Circular 7908. U.S. Bureau of Mines. Washington, D.C., 1959.

———. *Uranium Mining in the Lukachukai Mountains, Apache County, Arizona, Kerr-McGee Oil Industries, Inc.* U.S. Bureau of Mines, Washington, D.C., 1961.

Dare, Wilbert L., R. A. Lindblom, and J. H. Soule. *Uranium Mining on the Colorado Plateau.* Information Circular 7726. U.S. Bureau of Mines. Washington, D.C., September 1955.

Duncan, D. C., and W. L. Stokes. *Vanadium Deposits in the Carrizo Mountain District, Navajo Indian Reservation, Northeastern Arizona and Northwestern New Mexico.* Manhattan Engineering District, U.S. Army Corps of Engineers, U.S. Department of the Interior, U.S. Geological Survey. Washington, D.C., 1942.

Ford, Bacon and Davis Utah, Inc. *Engineering Assessment of Inactive Uranium Mill Tailings, Monument Valley Site, Monument Valley, Arizona.* Grand Junction, Colorado, March 31, 1977.

———. *Engineering Assessment of Inactive Uranium Mill Tailings, Shiprock Site, Shiprock, New Mexico.* Salt Lake City, Utah, July 1981.

Governors' Conference on Health Hazards in Uranium Mines: A Summary Report. U.S. Public Health Service Publication 843. Washington, D.C., 1961.

Hans, Joseph M., Jr., and Richard L. Douglas. *Radiation Survey of Dwellings in Cane Valley, Arizona, and Utah, for Use of Uranium Mill Tailings.* Environmental Protection Agency. Las Vegas, Nevada, August 1975.

Holaday, Duncan A. Draft testimony. *Radiation Exposures in Uranium Mines and Mills.* Joint Committee on Atomic Energy. U.S. Congress. Salt Lake City, Utah, February 19, 1959.

Holaday, Duncan A., Henry N. Doyle, and Wilfred D. David. *Interim Report of a Health Study of the Uranium Mines and Mills.* Federal Security Agency, U.S. Public Health Service and Colorado State Department of Public Health. Salt Lake City, Utah, May 1952.

Holaday, Duncan A., and Henry N. Doyle. *Progress Report (July 1950–December 1951) on the Health Study in the Uranium Mines and Mills.* U.S. Public Health Service. Salt Lake City, Utah, January 1952.

Holaday, Duncan A., David E. Rushing, Paul F. Woolrich, Howard L. Kusnetz, and William F. Bale. *Control of Radon Daughters in Uranium Mines and Calculations on Biologic Effects.* Public Health Service Publication 494. Washington, D.C., 1957.

Melancon, Paul E. "History of Exploration." *Geology and Technology of the Grants Uranium Region.* Edited by Vincent Kelley. Memoir 15. New Mexico Bureau of Mines and Mineral Resources. Socorro, New Mexico, 1963.

Office of Navajo Uranium Workers. *Results of the Uranium Miners Screening Project.* Shiprock, New Mexico, 1993.

Olsen, James H., Jr. *Jack Pile Reclamation Project—History and Progress Update*. Billings (Montana) Symposium, 1993.

O'Rear, Neilsen B. *Summary and Chronology of Domestic Uranium Program*. U.S. Atomic Energy Commission. Grand Junction, Colorado, 1966.

Pastore, U.S. Sen. John O. Press release. April 21, 1967.

Pond, Allen, Henry Doyle, John Johnson, Frank Costello, and Forrest Western. *Report of the Inter-Agency Committee on Health Hazards in Uranium Mining*. Washington, D.C., November 12, 1959.

Tamplin, Arthur R., and John W. Gofman. *The Colorado Plateau: Joachimsthal Revisited? An Analysis of the Lung Cancer Problem in Uranium and Hardrock Miners*. Presented as testimony. Joint Committee on Atomic Energy. 91st Cong., January 28, 1970.

Tome, Harry. Testimony. *Field Hearing of the U.S. Senate Committee on Labor and Human Resources*. Salt Lake City, Utah, April 14, 1982.

Udall, Stewart. Fact sheet. *Uranium Miners Claims Filed with the Department of Energy*. July 20, 1979.

———. "Newsletter to Navajo Uranium Miners and Widows." June 11, 1981.

———. Newsletter. April 12, 1982.

Uranium Study Advisory Committee. Meeting minutes. U.S. Public Health Service. Cincinnati, Ohio, February 26, 1953.

Uranium Study Advisory Committee. Meeting minutes. U.S. Public Health Service. Salt Lake City, Utah, December 1953.

U.S. Bureau of Mines. Department of the Interior. *Radiation Protection Activities in the Department of Interior*. Washington, D.C., August 9, 1960.

U.S. Department of Energy. *UMTRA Shiprock Completion Overview*. Shiprock, New Mexico, October 3, 1986.

U.S. Department of Energy. *Uranium Mill Tailings Remediation Action Project*, Shiprock, New Mexico, Groundwater Project, February 1993.

U.S. Department of Energy. *Uranium Mill Tailings Remedial Action Project, Tuba City, Arizona, Groundwater Project*. Albuquerque, New Mexico, March 1993.

U.S. Department of the Interior. *Environmental Issues and Uranium Development in the San Juan Basin Region*. Albuquerque, New Mexico, November 1979.

Wolf, Bernie, and Merril Eisenbud. *Medical Survey of Colorado Raw Materials Area*. U.S. Atomic Energy Commission. Washington, D.C., July 1948.

CONGRESSIONAL HEARINGS (chronological)

U.S. Senate Committee on Indian Affairs. *Leasing of Indian Lands*. 57th Cong., 1st sess., 1902.

U.S. Senate Subcommittee on Labor. Committee on Labor and Public Welfare, RE: H.R. 8989, S. 2972, S. 996, S. 3094. 89th Cong., 2nd sess., March 21, 22, 1966.

U.S. Congress. Subcommittee on Research, Development, and Radiation. Joint Committee on Atomic Energy. *Radiation Exposure of Uranium Miners*. 90th Cong., 1st sess., 1967.

U.S. Senate Special Committee on Aging. *Occupational Health Hazards of Older Workers in New Mexico*. Grants, New Mexico, August 30, 1979. 96th Cong., 1st sess., 1980.

U.S. Senate Subcommittee on Mineral Resources Development and Production. Committee on Energy and Natural Resources. *Impacts of Past Uranium Mining Practices*. Shiprock, New Mexico, March 13, 1990. 101st Cong., 2nd sess., 1990.

U.S. Senate Committee on Labor and Human Resources. *Examining the Implementation of the Radiation Exposure Compensation Act of 1990*. Shiprock, New Mexico. 103rd Cong., 1st sess., June 5, 1993.

ARTICLES: MAGAZINES AND JOURNALS

Anderson, H. L. *Bulletin of Atomic Scientists* (September 1974).

Jervy, Gay. "Stewart Udall's Newest Frontier." *American Lawyer* (January/February 1991): 72–81.

Lorenz, Egon. "Radioactivity and Lung Cancer: A Critical Review of Lung Cancer in the Mines of Schneeberg and Joachimsthal." *Journal of the National Cancer Institute,* 5, no. 1 (August 1944): 1–15.

Lundquist, A. Q., and J. L. Lake. "History and Trends of the Uranium Plant Flowsheet." *Mining Congress Journal* (November 1955): 37–42.

Shuey, Chris. "The Widows of Red Rock." *Saturday Magazine. Scottsdale Progress*, June 9, 1979, 5–7.

Smart, Dan. "Tragedy at Red Rock." *Oklahoma Observer*, February 25, 1979, 14–19.

Stewart, Ethelbert. "Survey of Industrial Poisoning from Radioactive Substances." *Labor Review* 28, no. 6 (June 1929): 20–65.

ARTICLES: NEWSPAPERS

"Aid for Navajo Miners Who Have Lung Cancer." *Navajo Times*, October 11, 1973.

Goldenberg, Bene. "Navajo Family Subsidies Sought, BIA Investigation Begins." *Albuquerque Tribune*, June 14, 1974.

Ivins, Molly. "Uranium Mines Leaving Indians a Legacy of Death." *New York Times*, May 20, 1979.

Juarez, Macario, Jr. "Radiation Compensation, U.S. Pays First Navajo Claims." *Gallup Independent*, May 28, 1992.

Lucas, Urith. "Navajos Who Mined Uranium Dying from Lung Cancer, Relatives Seeking Federal Compensation." *Albuquerque Tribune*, August 17, 1973.

"Lung Cancer Hearings for Workers in Uranium Mines Are Scheduled." *Navajo Times*, circa November 1973.

Madsen, Reed. "Utah Reluctant to Pay Mine Widows." *Deseret News*, June 6, 1979.

Reistrup, J. V. "Hidden Casualties of Atom Age Emerge." *Washington Post*, March 9, 1967.

Saltzstein, Kathie. "EPA Head Seeks Help for Red Valley Navajos Affected by Uranium." *Gallup Independent*, April 7, 1980.

Senia, Al. "Suit Says Indian Miners Were Exposed to Radiation." *Washington Post*, December 15, 1979.

Wheeler, Ellen. "$44 Million Urged for U-Miners, Kin." *Rocky Mountain News*, September 6, 1979.

White, Eliot Gordon. "Why Utah Fallout Aid Is Slow in Coming." *Deseret News*, July 20, 1979.

LETTERS, MEMORANDA

(All letters in private collections unless otherwise noted.)

Allpert, R. H. Letter to H. I. Smith, U.S. Geological Survey, December 14, 1949.

Archer, Victor E. Office memorandum. "Permission to Testify at Colorado Hearing, Regarding Medical History of Tom Van Arsdale." U.S. Public Health Service, October 10, 1957.

———. Letter to Stewart Udall, May 7, 1993.

Batie, Ralph V. Letter to Stewart Udall, December 26, 1979.

Domenici, U.S. Sen. Pete. Letter to Harry Tome, March 25, 1977.

Doyle, Henry N. Letter to Dr. Wilfred D. David, August 8, 1948.

———. "Survey of Uranium Mines on Navajo Reservation, November 14–17, 1949, January 11–12, 1950." U.S. Public Health Service. Salt Lake City, Utah, 1950.

Gerson, Stuart M. Letter to Stewart Udall, January 14, 1993.

Hansen, Eldon W. Letter to James Nahkai, Chairman of Budget and Finance Committee, Navajo Nation, April 15, 1981.

Holaday, Duncan. Letter to Allen D. Look, August 27, 1953.

———. Letter to Jim DeMoss, September 21, 1953.

———. Office memorandum, September 30, 1953.

———. Letter to Stewart Udall, January 29, 1979.

———. Letter to Stewart Udall, May 21, 1983.

———. Letter to Stewart Udall, January 16, 1984.

———. Letter to Stewart Udall, July 20, 1984.

Krider, Frank. Letter to Stewart Udall, April 7, 1993.

Lujan, U.S. Rep. Manuel, Jr. Letter to Harry Tome, March 24, 1977.

Mahoney, William P., Jr. Letter to Harry Tome, October 3, 1980.

Miller, Seward E. Office memorandum. "Pathological and Chemical Studies of Tissues from Autopsied Miners," February 17, 1954.

Runnels, U.S. Rep. Harold. Letter to Harry Tome, April 1, 1977.

Snapp, Roy B. "Radiation Problem in Uranium Mines of the Colorado Plateau." Atomic Energy Commission, April 4, 1952.

Strauss, Lewis L. Letter to W. Sterling Cole, Chairman, Joint Committee on Atomic Energy, July 13, 1953.

Tome, Harry. Letter to U.S. Rep. Harold Runnels, June 1975.

———. Letter to George Vlassis, General Counsel, Navajo Tribal Council, December 11, 1979.

Tome, Harry, and Elwood Tsosie. Letter to Stewart Udall, September 29, 1978.

Udall, Stewart. Letter to Harry Tome, December 15, 1978.

———. Letter to Harry Tome, July 3, 1979.

———. Letter to Joseph DiStefano, August 22, 1979.

———. Letter to Peter MacDonald, Chairman of Navajo Tribe, May 28, 1980.

———. Letter to Stuart Gerson, U.S. Assistant Attorney General, and Helene Goldberg, Director, Torts Branch, U.S. Department of Justice, December 17, 1992.

UNPUBLISHED MATERIAL

Chenoweth, William L. Oral History Project, 1990 (Grand Junction, Colorado) and personal communication, March 1993.

Red Rock Chapter, Navajo Tribe. Resolution RRC-48–76.

Shumway, Gary Lee. "A History of the Uranium Industry on the Colorado Plateau." Ph.D. diss., University of Southern California, 1970.

STATUTES

U.S. Statutes at Large 41 (1919): 3–34 [H.R. 2480]: Chapter 4. 66th Cong., 1st sess.

U.S. Statutes at Large 104 (1990): 920–26 [Public Law 101–426] Radiation Exposure Compensation Act, October 15, 1990.

Index

Note: Photo citations are in ***boldface****.*